W9-AMZ-900

WITHDRAWN

# New Methods in Symbolic Logic

GERALD B. STANDLEY
*University of Florida*

HOUGHTON MIFFLIN COMPANY · BOSTON
*New York · Atlanta · Geneva, Illinois · Dallas · Palo Alto*

To Dolores





Printed in the U.S.A.

Library of Congress Catalog Card Number: 76-140669

ISBN: 0-395-05429-X

# Preface

The student strong in mathematics very often has an advantage over other students not only in finding logic more accessible but also in encountering texts written from his point of view, so to speak. I have undertaken — I hope without prejudice to him — to write a text which will be equally clear to the student whose background in mathematics is not at all strong. In short, by using non-mathematical language, concrete examples, and frequent appeals to intuition, I have tried to insure that no student is at a disadvantage in approaching the discipline's cold but beautiful abstractions.

An introductory text in symbolic logic can follow either of two procedures. It can start with an axiomatic system and derive from it the methods of analyzing arguments; or it can begin with a body of problem-solving methods, some of which can be organized into a system derivable from axioms. Whichever procedure is elected, certain topics must be treated: notions of the abstractions and consequent power of the symbolism, rigor of proof, devices by which proofs are effected, the systematic features of consistency and completeness, and so on. Thus the real choice is which procedure will best develop the student's understanding. The present text is committed to the problem-solving approach, and the teacher or student to whom this appeals will find here a number of new methods purporting to exploit fully the advantages of this alternative. The methods are not introduced merely because they are new but because they seem to me to have proved their ability to acquaint beginners with logic. To the extent that the student understands the mechanization of logical techniques, he is given something more than a labor-saving device; he is afforded a better insight into the nature of formal logic. No method has been given a place here unless I have been satisfied that it serves this purpose.

The particular form given natural deduction in this text provides a rule of inference which allows certain qualifying components of an expression to be replaced by what they imply. This device provides even more flexibility in the quantificational logic than in the propositional. The cross-out technique is entirely effective in the propositional logic and in the quantificational comes closer to being effective than any other method I have been able to devise. An ideographic method using Parry's trapezoid symbol offers a streamlined substitute for truth tables and is also useful for singly quantified monadic functions.

Still another departure from conventional methods is the formulation of the rules for instantiation and generalization. Here the justification is almost purely pedagogical. Too often students learn to perform these operations by rule without understanding the reasons for the rules themselves. The present formulation and exposition are intended to show exactly what the rules are coping with. Section 6 of Chapter IV is necessarily somewhat difficult, but the student's reward for mastering it is that he will understand the essential nature of polyadic functions.

Chapter III, Class Logic, serves as a bridge between propositional and quantificational logic. This function justifies the explanation of existential import, the considerations in its fourth section, and the appearance of the Square of Opposition in Chapter IV instead of here. Having taken up class logic, I have tried to insure that none of its major features is omitted entirely. The reason such short shrift is given immediate inference and the methods of testing validity is that they are developed only so far as will help in intuiting the symbolism of quantification. The instructor who prefers to devote more time to class logic will discover, I believe, that the topics are suitably arranged for classroom elaboration. Those attempting to reach the end of the text in one semester may prefer to reduce assignments in this chapter or to skip it, referring to it only as Chapter IV may require.

Section 4 of Chapter II, on the simplification of DNFs, requires comment. The first six paragraphs are essential to the Deductive Method developed by the text and to subsequent theoretical discussion of it. But the remainder of the section and part of the exercise following it are difficult and excursive. This material is included so that, if time allows, the student may savor a problem to which logicians have devoted a great deal of thought and energy. For those interested in computers or circuit design, it offers methods that are new and extremely practical.

Professor Joel Friedman has helped immensely in the tiresome but indispensable process of freeing the text of defects, some of which were both egregious and consequential and might have found their way into print but for him. His help is here most gratefully acknowledged. I am also indebted to Professor William T. Parry for a number of valuable suggestions. For such faults as remain, the sole responsibility is my own.

*Gerald B. Standley*

# Contents

## III Class Logic



## IV Quantificational Logic



## V Further Aspects of Quantificational Logic

# I

# Propositional Symbolism

## 1. The Nature of Logic Today

A mid-nineteenth century logic course had as its purpose acquainting the student with the means of analyzing the syllogism. The tools he learned to use were forged by Aristotle better than two millennia earlier. Today's student of logic approaches a field as much more developed than this as calculus is higher than arithmetic. The earlier logic was fairly well contained by the question "What constitutes sound reasoning?" The new logic presses hard against the limits of such questions as "What are the characteristics of an axiomatic system? How can we be certain that an axiom is not really a theorem derivable from other axioms? Will a given group of axioms yield everything that can be said within their domain? How can we know they will never yield anything false?"

What is meant by *false* in this context is nothing more than *inconsistent*. The notion that a vast edifice of knowledge can be erected by starting with something certain and evolving only what is thereby implied has frustrated human thinking often enough that we now call such ideas naive. The *what is implied* part offers difficulties enough, but the *certainty* that must serve as the foundation is even more moot. Though it is an oversimplification, it is hardly an exaggeration to say that much of the course of modern philosophy, of which Descartes is justly looked upon as the father, has been determined by these two questions.

The logician, *qua* logician, has no part in these disputes. In respect to real life matters, he stands in much the same position as does the pure mathematician when bridge-building is being discussed. The mathematician stands ready with the abstract tools and formulas upon which the engineers' skills depend; the empirical questions as to the compression strength of the concrete, the tensile strength of the cables, or the solidity of the soil are matters in which he has no voice. Similarly, the logician cannot be asked to answer the question "What is certain?" or "What is true?" His concern is with an altogether abstract question: If the *postulates* be thus-and-so, what *follows* according to what agreed-upon *rules* of deduction?

This comparison of the logician and the pure mathematician is an apt one, for both may be thought of as working in complete independence of practical concerns. Of course, what the mathematician discovers almost always has its practical application, though sometimes only after many years. The logicians' findings have similarly been appealed to by sociologists, psychologists, and theologians as well as by philosophers. But the logician, like the pure mathematician,[1] is concerned with something wholly abstract. His study may be said to start and end with the blackboard. His symbolism may be given whatever interpretation its application requires; such interpretations and applications are, strictly speaking, of secondary importance to him. There is an element of exaggeration in this last statement. Logicians are frequently concerned that a system be applied to human discourse. But more and more as time goes on, logicians contemplate systems either not immediately applicable or else applicable only to the equally blackboard-limited realm of pure mathematics.

Logic and mathematics have grown increasingly closer since the middle of the eighteen hundreds. The long inquiry into Euclid's fifth postulate ultimately produced not only the non-Euclidean geometries but also a wholesome realization that the 'science' of mathematics rests only on so many presuppositions. Inquiry in the last hundred years has confirmed the suspicion that presuppositions are only presuppositions no matter how infallible they may appear, and this inquiry has driven mathematics ever nearer to logic. Logic — or at least what is called modern logic — at last cut the umbilical cord that bound it to language and with the symbolism developed by Venn, Boole, Frege, Peano, Russell, Whitehead, and many others has been able to reach into the question of what presuppositions underlie mathematics in its various branches. This inquiry, of course, has driven logic ever nearer to mathematics. The term *mathematical logic*, as used during the first quarter of the twentieth century, referred principally to the fact that logic, like mathematics, had embraced an abstracted symbolism. During the second quarter of the century the implications of this fact flowered forth in an entirely different meaning of the same phrase; today mathematical logic means logic concerned with mathematics.

As a result, the logic course of today undertakes first to acquaint the student with its symbolism, for this is the means for approaching the problems currently of greatest importance in this field.

[1]Notice that we need never say *the pure logician* since there is no other kind. The thinker who appeals to logic to support this or that argument is dubbed sociologist, psychologist, theologian, philosopher, etc.

## 2. The Symbols

Symbolism comes about from the logicians' concern not with the *content* of deductive reasoning but with its *form*. It is true that the beginner in logic spends a good deal of time testing arguments for their validity and that these arguments are frequently expressed first in ordinary English and so have a particular content, *i.e.*, premises and conclusions that actually say something. But such arguments are analyzed by their form, and their content is commonly inconsequential or even frivolous. More than this, they are often arguments which can be easily intuited as valid or invalid, so that the real purpose of the exercises in a logic book like this one is to perfect the student's mastery of the tools of analysis. There is paradoxical evidence of this. The problems in ordinary language, if they occur at the beginning of the book where our incursions into symbolism are still rudimentary, may indeed offer some resistance to intuition; but the problems in the later chapters, where our symbolism has become more complex and detailed, are sometimes so simple that a child can be sure they are valid. The symbolism of logic can be rightly construed as a determination to reproduce as mechanically as possible the deductive thinking that takes place in our brains in a split second without our knowing how. To the extent that the symbolism succeeds in entering into the detail of such reasoning, attention is given to simpler examples. And this is just as natural as it is for one who undertakes to construct extremely accurate scales to weigh minute things. Accordingly, it is with form that logic is concerned. The content of any argument is nothing more than an inconsequential vehicle of its form. And symbolism is the device for distilling the essence.

It is to circumvent considerations of content that we say, "Let this proposition (*Aunt Charlotte is singing*, *Chicago is west of Memphis*, *The Portuguese defend their colonialism*, or what-have-you) be represented by $p$, let this other proposition be represented by $q$, this third one by $r$, etc." And if there appears a contradiction of the previously encountered proposition, *e.g.*, *Aunt Charlotte is not singing*, this is represented by the same letter, $p$, but with the difference that $p$ is now denied. A bar over the letter is the convention for representing this fact. Thus if

'Aunt Charlotte is singing' is represented by $p$,

'Aunt Charlotte is not singing' is represented by $\bar{p}$ (read *non-p*)

This bar is an operator. Its presence over a letter has the effect of contradiction, *i.e.*, $p$ is contradicted by $\bar{p}$, and vice versa. Letters so used are called propositional *variables* inasmuch as they can be assigned various contents. It is true that throughout any particular argument $p$ must always stand for the same proposition, but once that argument is set aside and the next one taken up, $p$ may be assigned an entirely different content. An operator, on the other hand, is a *constant*. Whatever $p$ may mean, $\bar{p}$ has a contradictory meaning; $\bar{p}$ may be said to negate $p$ and vice versa, whatever the context.

Whereas negation always operates thus on a single proposition, there are other operators that place two propositions in some sort of relationship to each other. This is technical language for a common enough idea. When it is affirmed that Aunt Charlotte is singing and Uncle Jim is shutting his door, we are putting two propositions into a relationship with each other. The name of this relation is *conjunction*, and it is symbolized by a dot placed between the variables, though sometimes it will be convenient to omit the dot, allowing the mere juxtaposition of the two variables to signify that they are conjoined.

> 'Aunt Charlotte is singing and Uncle Jim is shutting the door': $p \cdot q$ or $pq$ (read *p and q* or *pq*)

Our ordinary language provides another logical operator called *disjunction* and represented by a wedge (or *vel*):

> 'Either Aunt Charlotte is singing or Uncle Jim is shutting his door': $p \vee q$ (read *p or q*)

Still another operator is called *material implication* and uses a horseshoe symbol. It is read *if p then q* or *p implies q*. The symbolism $p \supset q$ would in the present context be translated back into English as "If Aunt Charlotte is singing, then Uncle Jim is shutting the door."

Such compound statements can be distinguished from their parts by remarking that the variables are symbols of content whereas the operators are symbols of form first and then of content. In a compound statement such as $p \supset q$ the content varies; the form is constant. Another important feature is that by establishing a relation between the two variables a new but dependent affirmation emerges. The statement $pq$ is a new statement in that it affirms that $p$ is true and that $q$ is also true. This new assertion would be called false despite the truth of one of the component propositions if the other were not also true. For this reason, although the affirmation that both are true is *new*, it is still *dependent* on its constituent parts.

An unexcelled means of clarifying this dependence is the truth table, a logical device brought into use in the 1920's and so called because it tabulates systematically all the combinations of truth and falsity possible in the component variables. These combinations can be set down with the letters *T* for true and *F* for false, or, following the binary system of numeration which has found so many applications in the electronic age and which is obviously well adapted for the present purpose, truth can be represented by *1* and falsity by *0*. With two variables there are four such combinations.

| $q$ | $p$ |
|---|---|
| 1 | 1 |
| 1 | 0 |
| 0 | 1 |
| 0 | 0 |

At first glance it may appear strange that $p$ should head the right-hand column of this table. This strangeness will be the greater for those already introduced to the truth table, for the common usage is the reverse of this. But the present departure is anything but capricious. Any numerical table that is ordered (and this one is in descending order) must grow from right to left as it is expanded. By placing the first variable, $p$, over the unit column of the numeration, $q$ over the second column from the right, $r$ over the third column, and so on, the addition of further variables is made to accompany the numerical growth of the table, with the result that the table for $n$ variables remains unchanged, except for duplication, when the table is expanded to $n + 1$ variables.

| $r$ | $q$ | $p$ |
|---|---|---|
| 1 | 1 | 1 |
| 1 | 1 | 0 |
| 1 | 0 | 1 |
| 1 | 0 | 0 |
| 0 | 1 | 1 |
| 0 | 1 | 0 |
| 0 | 0 | 1 |
| 0 | 0 | 0 |

This characteristic, which may here seem nothing more than a minor elegance, will expedite enormously the development of the trapezoid method in Chapter II.

Having once tabulated the possible combinations of the truth and falsity of $p$ and $q$, the precise meaning of the new statement $p \cdot q$ can be entered in the table. It is true when both of the component variables are true, and false under all other conditions.

| $q$ | $p$ | $p \cdot q$ |
|---|---|---|
| 1 | 1 | 1 |
| 1 | 0 | 0 |
| 0 | 1 | 0 |
| 0 | 0 | 0 |

Logicians were rightly entranced by this purely formal way of defining the meaning of conjunction. Among other things, it removes every vestige of ambiguity that invests a natural language such as the English we ordinarily communicate in. The word *or* serves as a good example of such ambiguity. When a hostess offers her guest coffee or tea, she would be surprised to hear him reply "Both, please" although the same reply to "Sugar or cream?" would be quite natural. This is because the *or* sometimes means *one or the other but not both;* at other times, *one or the other or both.* Frequently, as in the example given, the context provides the interpretation. When cast into symbols, the meaning of *or* cannot be left in doubt. From the following truth table for the wedge, one can readily discern whether that symbol is to stand for the 'exclusive' *or* — coffee or tea — or the 'inclusive' *or* — sugar or cream.

| $q$ | $p$ | $p \vee q$ |
|---|---|---|
| 1 | 1 | 1 |
| 1 | 0 | 1 |
| 0 | 1 | 1 |
| 0 | 0 | 0 |

**EXERCISE I–2**

1. Write these in symbols. Use '$\therefore$' as a symbol for *therefore*. (Answers to starred question will be found at the end of the book.)
   a. Aunt Charlotte is singing and Uncle Jim is shutting the door.
   b. Sally has gone out.
   c. Therefore Uncle Jim is shutting the door and Sally has gone out.
   d. Sally has not gone out or Aunt Charlotte is not singing.
   e. If Uncle Jim is not shutting his door, then Sally has gone out.
   f. Either Uncle Jim is shutting the door or Aunt Charlotte is singing or Sally has gone out.
   g. Uncle Jim is shutting his door and Sally has not gone out.
   *h. Uncle Jim is shutting his door but Sally has not gone out.
2. Write English sentences for which the following would be appropriate symbols.

   a. $p\bar{q}$ b. $p \supset \bar{q}$ c. $\therefore \bar{q} \vee \bar{p}$ d. $\bar{p}qr$
3. Make a truth table for the exclusive *or*, using a broken wedge to symbolize it: $p\backslash\ /q$. Examine each horizontal line of the table to determine whether a *1* or a *0* should be entered under this operator.

### 3. The Paradoxical Material Implication

From the remarks in the previous sections it is clear that it would be perfectly feasible to assign meanings to arbitrarily made-up operators having no easily stated counterparts in ordinary language. For example, Sheffer's 'stroke' function (or operator) can be given a truth table definition very easily.

| $q$ | $p$ | $p \mid q$ |
|---|---|---|
| 1 | 1 | 0 |
| 1 | 0 | 1 |
| 0 | 1 | 1 |
| 0 | 0 | 1 |

What it means is that at least one of the statements it connects is false. This relation is not ordinarily used in speaking; were we to relate *Aunt Charlotte is singing* and *Uncle Jim is shutting the door* in the way depicted by this truth table, we might well say, "Either Aunt Charlotte is not singing or Uncle Jim is not shutting the door" which has recourse to disjunction and negation and so might be as readily rendered $\bar{p} \vee \bar{q}$, which is logically the equivalent but does no better than the previous sentence in avoiding recourse to negation. Indeed, we have no linguistic counterpart to $p \mid q$ as we do to $\bar{p}$, or to $p \vee q$, or $p \cdot q$.[2] Logicians have had occasion to use it despite this, however, since it affords a formal relation very useful for certain systematic purposes.

With this understood, the student is prepared to give further consideration to material implication, frequently a stumbling block for beginners in logic. Surprising though it may be, logicians have no handy way of stating that Uncle Jim is shutting his door *because* Aunt Charlotte is singing. True-false logic gets ahead swimmingly despite this, using an implication (called *material* to acknowledge its oddness) which is defined by this truth table.

| $q$ | $p$ | $p \supset q$ |
|---|---|---|
| 1 | 1 | 1 |
| 1 | 0 | 1 |
| 0 | 1 | 0 |
| 0 | 0 | 1 |

The disturbing feature of this definition is that it seems to say that a false statement implies a true one (line 2 of the table) and even that a false statement implies a false one (line 4). What is even worse, we are committed to calling $p \supset q$ true when the antecedent $p$ may have nothing whatever to do with the consequent $q$, *e.g.*, '*New York is larger than Chicago* implies that *the first Americans were Indians*' becomes a true statement (line 1)! The easiest way to encompass these difficulties is to remind oneself that this is *material* implication and so pass over any artificiality out of consideration of its usefulness. We dare to cast what we ordinarily mean by implication into

[2]*Not both p and q* (*i.e.*, "It is not true both that Aunt Charlotte is singing and Uncle Jim is shutting the door") is another rendering, but again this reflects more perfectly the symbolism $\sim(p \cdot q)$ which will be seen shortly to be a negation of the conjunction of $p$ and $q$. Russell gives *p is incompatible with q*, which seems to be as close to a linguistic counterpart as we are likely to come. But because we do not ordinarily think of two false statements as incompatibles, even this rendering is defective with respect to the fourth line of the truth table unless we elucidate incompatibility either by one of the phrases already referred to or

the mold of material implication for two reasons: material implication is the best mold available short of modal logics,[3] and the procedure pans out surprisingly well after all.

Some consolation is available from an unexpected quarter. Actually, this strange relation occasionally appears in ordinary discourse. When one remarks, "If he can afford that car, my name is Croesus!" he is using the language of material implication. The meaning is unambiguous and is to be found on line 4 of the table. Since the implication is affirmed, there are only three lines, 1, 2, and 4, to choose from as it is only on these lines that the implication is true. Since the consequent is patently false, line 4 is the only meaning that can be intended. The assertion is an indirect but forceful way of saying the antecedent is false, *i.e.*, that *he certainly cannot afford that car.*

## 4. Material Equivalence and Compound Statements

The adjective *material* (which will soon be omitted) also attaches to the word equivalence. The truth table definition of $p \equiv q$ (read *p is equivalent to q*) is this:

| $q$ | $p$ | $p \equiv q$ |
|---|---|---|
| 1 | 1 | 1 |
| 1 | 0 | 0 |
| 0 | 1 | 0 |
| 0 | 0 | 1 |

It will now be no surprise that the statements *New York is larger than Chicago* and *the first Americans were Indians* are materially equivalent since all that is meant by this is that either they are both true or they are both false. *Al Capone was morally upright* and *Theodore Roosevelt was meek* are also materially equivalent (line 4). What is the significance of this for the abstract use of propositions? Since the emphasis in logic is always on form, once a proposition is abstracted from its content, all that can remain to it is to be found in the truth table, *i.e.*, all that matters is whether the proposition be true or false. Put more exactly yet, it makes little difference which of *these* it is in the present case; what matters most, of course, is that if the two statements are materially equivalent, one can be used just as well as the other.

---

as meaning *If p be true then q is false and if q be true then p is false* (relying, moreover, on material implication in construing the *if-then*). In short, unless we define $p \mid q$ by recourse to the truth table, we seem obliged to define it in terms of other operators more immediately reflected by language.

[3]The previous mention of true-false logic refers to logical systems like those being studied here in which values of true and false are the only ones assignable to a proposition. Multi-valued logics of three or more possible values also exist. Modal logic takes into consideration the *necessity* or the *possibility* of a statement's truth. About the time of the First World War, Lewis undertook the development of what is called *strict* implication with the help of these modes. The interested reader should consult *Symbolic Logic*, C. I. Lewis and C. H. Langford (New York: The Century Co., 1932).

This can be shown by examining the truth table for the expression $\bar{p} \vee q$.

| $q$ | $p$ | $\bar{p}$ (2) | $\vee$ (1) | $q$ |
|---|---|---|---|---|
| 1 | 1 | 0 | 1 | |
| 1 | 0 | 1 | 1 | |
| 0 | 1 | 0 | 0 | |
| 0 | 0 | 1 | 1 | |

To fill in the column under the wedge, it is first necessary to show the value on each line for $\bar{p}$. The column indexed with a *1* is the last to be filled in. Notice that this expression is false only on the third line of the table. Note further that the same thing is true of the expression $p \supset q$ (page 7). Since one expression is true under precisely the same assignments of truth or falsity to $p$ and $q$ as the other, and both are false under like conditions, it can be concluded that the two expressions are equivalent. This discovery can be set down thus:

$$(\bar{p} \vee q) \equiv (p \supset q)$$

There are two apparently trivial, but really quite important, differences between this expression and any previous one. One is that this is the first in which a statement about two statements appears ($\bar{p} \vee q$ and $p \supset q$), each of which is already about two statements (the component variables). The important thing here is that it is still *one* statement, now to the effect that the one compound statement is equivalent to the other. Once statements about statements about statements are admitted, there will obviously be no limit to the complexity of statements so compounded. This statement could itself become a constituent part of a larger one, and that part of a larger one yet, etc. It will always be true, however, that no matter how long such an expression may become it will always comprise just one statement. The next section will have more to add on this very important point.

The other noteworthy feature of the new statement is that it is always true. If, guided by the definition of the symbol '$\equiv$', one checks each line of the truth table he will discover that this is the case. This *truth under all conditions* is the characteristic of the statements known as *tautologies*; these are understandably said to be *logically* true. A *self-contradiction*, by the same token, is a statement which is logically false, *i.e.*, false on each line of the truth table. Only one category remains: those statements which are true under some conditions (*i.e.*, on some lines) and false under others are called *contingencies*, their truth or falsity being contingent on, rather than independent of, the truth values of the component variables.

A material equivalence is sometimes called a *biconditional. If she loves him she will marry him* can be symbolized as $p \supset q$. *Only if she loves him will she marry him* can be written — keeping the same variables for the statements — as $q \supset p$. *If and only if she loves him will she marry him* can be written either as $(p \supset q) \cdot (q \supset p)$ or as $p \equiv q$. In short, to say that two statements imply each other is to say they are equivalent.

Of Philo is not.
The man went who is a was.
Wherever she spoke the audience afterwards.
Were twice him.

No speaker of English would confuse the above expressions with declarative sentences. Too many or too few nouns or verbs can render a series of words ungrammatical and ill-formed as far as accepted usage goes. If this be true in a natural language such as English, it is *a fortiori* the case with an artificial language designed from its inception to be unambiguous.

Here is an ill-formed (therefore pseudo-) statement in the propositional symbolism:

$$p \cdot q \vee r$$

At this point there is no way of telling what statement is intended. Does it mean $p \cdot (q \vee r)$ or $(p \cdot q) \vee r$? Note that each of these last two statements is well-formed. The first is the conjunction of $p$ with the disjunction of $q$ and $r$. In this the dot is the major operator, and the wedge, enclosed in parentheses, is one of the components of the conjunction; the wedge is subordinate to, or simply 'under', the dot. In the statement to the right, the wedge is major and the dot is under it. It is a disjunction between the conjunction of $p$ and $q$ as one member, and $r$ as the other.

This illustrates one of the requirements a compound statement must meet if it is to be well-formed. The two requirements can conveniently be outlined like this:

I. Every operator has the proper number of component statements in the appropriate places.

A. Negation is a singulary operator. The appropriate place for its single component is:

1. beneath the operator if the component is a variable (the operator is in this case a bar). Example: $\overline{p}$

2. enclosed in parentheses to the right of the operator if the component is a compounded statement (the operator in this case is a curl, or tilde). Example: $\sim(p \vee q)$

So far there has been occasion to negate only variables, as in $\overline{p}$, $\overline{q}$, etc. But of course some symbol is required for negating a statement constituted by an operator. The term *negation* will apply equally well to either symbol. Because it will be very convenient later to be able to distinguish between negations of variables only (the bar) and those of disjunctions, conjunctions, etc. (the curl), these two more specific names of symbols will remain in use.

B. The dot, wedge, horseshoe, equivalence, broken wedge, and others are binary operators and so establish a relation between two statements. The appropriate places for statements so related are on either side of the function.[4]

II. As a consequence of the first requirement, every operator is either itself major or subordinate immediately to only one operator. Hence any operators present in a well-formed statement are ordered as to their ranking.[5]

In $(\bar{p} \vee q) \cdot r$ the conjunction is major, the negation of $p$ is immediately subordinate to the wedge, which is, in turn, immediately subordinate to the dot. The defect of $p \cdot q \vee r$ was that it lacked this ranking of operators, a defect remedied in one way by $p \cdot (q \vee r)$ and in another by $(p \cdot q) \vee r$.

To all purposes there is neither temptation nor occasion to mistake as well-formed any expression failing to meet the first of these two requirements. Symbols such as $q\vee$, $\supset\equiv$, or $\vee\sim\vee$ are plainly ill-formed. But a careless failure to meet the second condition — that of ordering the operators in an expression — can easily result in a pseudo-statement. This is the obvious connection, then, between beneformation (the state or property of being well-formed) and punctuation (whatever device is used to establish this ordering of the operators).

There are several such devices. The one just used in which parentheses enclosing statements serve to subordinate them to operators lying outside those parentheses is only one. Another simple means of ordering the operators is to key them with numerals, using *1* over the major operator, *2* over the operators constituting statements subordinate thereto, *3* over the next in the hierarchy, etc. This is very helpful in working out truth tables. Thus the expression

$$(\bar{p} \vee \sim(q \cdot r)) \supset (s \supset t)$$

might be written thus:

$$\overset{3}{\bar{p}} \overset{2}{\vee} \overset{3}{\sim} q \overset{4}{\cdot} r \overset{1}{\supset} s \overset{2}{\supset} t$$

[4]Operators for more than two components exist, but these can always be replaced by more common operators and need not be included here. For example, $\mu$ can be allowed to mean that at least half of its components are true. If two components are assigned this function, it means exactly what the wedge means; if it relates three statements, $(\mu pqr)$ it means $pq \vee (qr \vee pr)$. This 'majority' operator can accordingly have any plural number of components.

[5]In the exercise expressions will come to light in which the ordering of the operators is indifferent, as in arithmetic it makes no difference whether $2 + 4 + 5$ is interpreted as $(2 + 4) + 5$ or as $2 + (4 + 5)$. Such *associative* operators need not be construed as unordered, even when written $2 + 4 + 5$; they can equally well be thought of as ordered in any of two or more indifferent ways.

The word *subordinate* is here qualified by *immediately* because an operator can be said to be subordinate to all operators ranking higher in the ordering. Thus in $(\bar{p} \vee q) \cdot r$ the bar stands under the wedge and also under the dot, but it stands immediately under the wedge only.

Lukasiewicz devised an elegant notation in which the operators are assigned their order by the very order of their writing. He would write the above expression as

$$\mathrm{CAN}p\mathrm{NK}qr\mathrm{C}st$$

an array less formidable to the practiced than the unpracticed eye. It affirms an implication (C) between the disjunction (A for *alternation*, another name for disjunction) between the negation (N) of $p$ and the negation (N) of the conjunction (K) of $q$ and $r$, and the implication (C) between $s$ and $t$. By writing the numbers of the numbered notation over the corresponding operators of this one, it will be observed that the appropriate places for an operator's components are to its right, and that there is here the proper number of such components.

$$\begin{array}{l} 1\,2\,3\;\;3\,4\;\;\;2 \\ \mathrm{CAN}p\mathrm{NK}qr\mathrm{C}st \end{array}$$

Every variable is immediately subordinate to the nearest operator to its left — accordingly the last C and the K have the proper number of components. Each negation has but one immediately subordinate statement — that immediately to its right. Immediately subordinate to the first A are the two *3*'s to its right. Subordinate to the first C are the two *2*'s to its right. Notice that the second *2* is not written until the compounded statement comprised by the first *2* is set down in its entirety. Lukasiewicz' notation would come to mind were one looking for a notation for programming a logic machine. It has also been used in the design of a game called *WFF 'n PROOF* (the *WFF* of which stands for *well-formed formulas*) in which the players cast dice bearing on their faces small letters (for the variables) and capitals (for the operators) which are then arranged *à la* Lukasiewicz into statements.

Mastering any form of punctuation entails some effort, the reward being the ability to perceive at a glance the gist of even a very complex expression. Perhaps the highest ratio of reward to effort obtains in what is called the dot system. In this notation dots are set beside the binary operators (on either side); the more important the operator, the more dots it receives. With a little practice the eye can see which statements are subject to which operators as easily as with the parentheses notation; the advantage is that the confusing cluster of vertical markings that characterize the parentheses notation is obviated. This notation skillfully exploits the option of writing conjunctions with or without dots:

$p \cdot (q \vee r)$ is written $p \cdot q \vee r$
$(p \cdot q) \vee r$ is written $pq \vee r$

This convention governs this notation: a conjunction in which the dot is left unwritten is subordinate to any other operator, even one without a dot; a conjunction in which the dot appears is more important than an undotted operator but less important than a dotted one. The hierarchy, in ascending order of importance, is thereby:

the unwritten conjunction
the dotless operator (other than conjunction)
the dot as conjunction
any (other) operator having a dot
two dots used to conjoin
any (other) operator having two dots
three dots used to conjoin
etc.

This hierarchy can easily be kept in mind with this simple rule: count each operator other than conjunction as a half-dot. This orders the above lines by construing them as having, successively, no dots, half a dot, one dot, a dot and a half, two, etc. Here is the expression from the previous page:

$$\bar{p} \vee \sim(qr) \,.\!\supset s \supset t$$

Note that the parentheses are preserved only in connection with the curl in order to show exactly what is being negated. The parentheses will always enclose some compound statement, of course, since a single variable is negated simply by placing a bar over it. In neither form does negation require any dots.

Beneformation can now be formally specified, even though this specification will anticipate some of the discoveries the student will make in doing the exercise at the end of the chapter. Let the word *literal* signify either a variable or a variable with a bar. Thus $p$ is a variable, but both $p$ and $\bar{p}$ are literals. Let the binary operators now be listed in such a way as to include any dot punctuation that may be attached to them:

1. Conjunction: any number of dots only, occurring together. (The special case of zero dots constitutes conjunction by juxtaposition.)
2. Disjunction: a wedge together with any number of dots. (Such dots are not conjunctions.)
3. Implication: a horseshoe together with any number of dots. (Such dots are not conjunctions.)
4. Equivalence: three bars together with any number of dots. (Such dots are not conjunctions.)

An expression is well-formed if and only if it qualifies under one of the following four heads:

1. It consists of a literal standing alone.
2. It consists of a literal followed by an even number of alternating binary operators and literals (in that order[6]), no two of these

[6]Juxtaposed literals or literals juxtaposed to expressions qualifying as well-formed under 3 are construed as being conjoined by a conjunction of zero dots.

binary operators to have the same number of dots unless

a. at least one of them is a conjunction, or
b. both are disjunctions, or
c. both are equivalences, or
d. a binary operator having more dots occurs somewhere between them.

3. It consists of a curl followed by a parenthesis containing an expression qualifying under 2 or 4.
4. It qualifies under 2, provided each expression qualifying under 3 is construed as a literal and its parenthesized content disregarded.

## 6. A Note on Double Negation

A unique feature of the symbolism legitimated by the above definition of beneformation is that a double negation is illicit. For example, $\sim(\bar{p})$ is malformed since the parenthesis contains no binary operator; $\sim(\sim(pq))$ is also ill-formed (the content of the outer parenthesis qualifies as well-formed under 3 whereas it would have to qualify under 2 or 4 for the entire expression to be well-formed). From this it might appear that there is no way of expressing *non-non-p*. But the concept of negation was introduced (page 3) by stating that $\bar{p}$ is the contradiction of $p$, and vice versa. It follows that *non-non-p*, being the contradiction of *non-p*, is simply written as $p$ without further ado. By so defining negation, no proof is required that the negation of $\bar{p}$ is $p$, nor need any rule be adduced for transforming *non-non-p* into $p$ — not even in a text devoted to methods — since *non-non-p* cannot occur within the symbolic language except as $p$. The present outlawing of doubly negated expressions, therefore, far from sidestepping any difficulty about double negation, obviates it as an unnecessary complication.

## EXERCISE I–6–A

Part of this exercise introduces essential material not to be found in the text proper.

1. Here is an expression punctuated with numerals. Is it the same expression as that on page 9?

$$\overset{3}{\bar{p}} \overset{2}{\vee} q \overset{1}{\equiv} p \overset{2}{\supset} q$$

2. Following the suggestion at the end of page 11, fill in the truth table for the above expression.

*3. What connection exists between numerical punctuation and the procedure of filling in a truth table?

4. Express symbolically this statement: The conjunction of $p$ with the disjunction of $q$ and $r$ is equivalent to the disjunction between the conjunction of $p$ with $q$ on the one side and $r$ on the other.

5. a. How many lines will a truth table for four variables have? For five variables?

   *b. What number of lines is required in the truth table for $n$ variables?

6. Using a truth table, determine whether the statement in question 4 is tautological, self-contradictory, or contingent.

7. Now that the dot punctuation is adopted, what does the previously ill-formed expression on page 10 mean?

8. Is $p \supset . q \supset r$ the same thing as $p \supset q . \supset r$? Do not rely on your intuition only, but use a truth table to substantiate your answer.

9. Is $p \vee . q \vee r$ the same as $p \vee q . \vee r$?

10. a. Is $p \vee q \vee r$ ambiguous? Well-formed?

    b. Is $pqr$ well-formed?

11. From your answers to questions 9 and 10 it should be apparent that disjunction and conjunction are associative. Show by truth tables that each of these operators is commutative, *i.e.*, that it makes no difference whether we write $p \vee q$ or $q \vee p$.

12. a. Is equivalence commutative?

    b. Is implication?

    *c. Which, if either, is associative?

13. a. Are there any symbolized expressions so far in this exercise which are ill-formed according to the formal definition of beneformation?

    b. Which of the following are ill-formed and in what respect do they fail to conform to the definition?

    1. $\sim(r \vee q$
    2. $\sim(\ )$
    3. $pr \sim (q \vee : r)$
    4. $p :: r$
    5. $p \cdot q \vee . r$
    6. $p \supset \sim(q \vee \bar{r} \vee \bar{s})$
    7. $q \equiv \sim(p \vee \sim(q \supset r))$

    c. Which, if any, of the well-formed expressions above can you write with fewer dots?

    *d. The definition of beneformation specifies (under heading 3) that the contents of the parenthesis is to be well-formed according to heading 2 or 4. In heading 4, reference is made to heading 3. Adams says this makes applying the definition to expression 7 above circular. Brown says the application is iterative instead. With which do you agree?

### EXERCISE I–6–B

1. Symbolize these:
   - a. Either the train is late or it won't come at all.
   - b. If Washington had accepted a third term, the tradition would have become common.
   - c. The tradition has never become common.
   - d. Therefore, Washington did not accept a third term.
   - e. If they had complained less or suffered more, they could be more easily forgiven.
   - f. If the marines had not been sent, our prestige would have suffered less but there would have been more bloodshed.
   - g. If it doesn't rain, it will stay hot.
   - *h. It will stay hot if it doesn't rain.
   - *j. It will not stay hot, provided it rains.
   - *k. If the river widens much at its mouth, we shall be in shallower water provided the current doesn't diminish.
   - *l. To say that Barclay will win if he carries the rural vote amounts to saying that either he wins or he fails to carry the rural vote.
   - *m. Barclay will win if he carries the rural vote if and only if he either wins or fails to carry the rural vote.
   - *n. Because John is sick, he won't be here.
   - *o. If John is sick, he won't come for that reason.
   - *p. Frank is willing and he is able.
   - *q. Frank is not both willing and able.
   - *r. Frank is unwilling and unable. (Note the distinction between this and the previous statement.)

2. Using the propositions given, write out the English equivalent for each of the symbolized expressions.

| | | | |
|---|---|---|---|
| $p$ | She is pleased | $t$ | She takes her time |
| $q$ | She quenches her thirst | $u$ | She understands little that is said |
| $r$ | She roars with laughter | $v$ | She voices her mind |
| $s$ | She sulks | | |

| | | | |
|---|---|---|---|
| a. | $pq \supset r$ | *e. | $\bar{r} \equiv . \, s \vee \sim(pt)$ |
| b. | $u \supset . \, v \supset s$ | f. | $\bar{r} \equiv s \, . \vee \sim(pt)$ |
| *c. | $rv \equiv u$ | g. | $\sim(pr) \equiv . \, \bar{p} \vee \bar{r}$ |
| *d. | $\bar{p}\bar{t} \supset \bar{v}$ | *h. | $u \vee pr \supset : t \equiv . \, s \vee q$ |

## SUMMARY

The introduction of symbolism in the latter half of the nineteenth century modified the nature of logic by expanding it. Today logic and mathematics tend to merge in their higher reaches. Mastery of the symbolism is a *sine qua non* of the study of logic.

Letters of the alphabet beginning at $p$ are widely used to represent propositions; they are called *propositional variables*. The constants, or operators, introduced in this chapter that will be in permanent use hereafter are negation, conjunction, disjunction, material implication, and material equivalence. The paradoxes attending material implication evanesce when it is rightly understood.

Every statement in the propositional logic is either a tautology, a self-contradiction, or a contingency.

Because ambiguous statements are never well-formed, the scope of each operator must be perfectly plain. All but the simplest well-formed statements rely on punctuation for their beneformation. Dot punctuation will be used in this text.

# II

# Methods in the Propositional Logic

## 1. Validity and Tautologous Form

Consider this simple syllogism:

If it rains, the game will be called off.
If the game is called off, we can't get our tenth victory today.
Therefore, if it rains, we can't get our tenth victory today.

There is no difficulty in intuiting that the argument is valid. If the first two statements, the premises, be true, the conclusion simply *must* be true, and this is precisely what is meant by validity. Even conclusions that are false can be validly reached, provided they *would be true* if the premises were true. To say that an argument is valid, then, is to say that the premises logically imply the conclusion.

By abstracting from the content, another more general syllogism can be discovered. Without its making any difference which statements $p$, $q$, and $r$ are made to stand for, the following syllogism is also valid:

$$p \supset q$$
$$q \supset r$$
$$\therefore p \supset r$$

Notice further that the language mastered in the first chapter can be used to state that if the premises are true, the conclusion must be.

$$p \supset q \cdot q \supset r . \supset p \supset r$$

Here, in its most general form, is the gist of the intuition by which the first syllogism is recognized to be valid. For if one were asked to say *why* the first syllogism is valid, he might well answer that if one thing implies a second, and that second a third, then the first certainly implies the third.

But this way of putting it is not merely more general. It can be tested by the truth table. By this means the general statement can be shown to be tautologous. This latter fact — that the statement is infallibly the case no matter what values are assigned to the component parts — is a stronger evidence that the argument is to be trusted than any ever-so-clear intuition can be. It is a way of showing not only *that* the argument is trustworthy, but *why* it is so.

Consider still further this important distinction between form and content. Suppose it turns out that there is no rain today. Logicians and semanticists have struggled with the resulting perplexing case (called the contrary-to-fact conditional). What is now the truth status of the first premise? If it fails to rain, how shall we ever know that the statement was true? Perhaps the umpire would have determinedly refused to call the game. Perhaps! This question is one of content. It never arises unless, as a matter of *fact*, the statement *it rains* is false; and even then we are inquiring into the factual truth of the implication expressed in the premise. Contrast with this the *formal* truth of that premise. By the definition of material implication, how should an implication whose antecedent is false be classified? But the matter of form goes still deeper. The truth table discloses that one statement is never anything except true: the *general form* symbolized above.

The reasons for calling the original syllogism valid are now clear. It is one specific *instance of a general syllogism which must always be valid.* And now to generalize, in its turn, *this* principle. This time $P$ is made to stand for the first premise,[1] $Q$ the second if there be a second, $R$ the third, and so on. Let $C$ represent in the same way the conclusion. It is apparent that if $PQRS \ldots \supset C$ be a tautology, the argument's validity is established. Thus symbolism provides a powerful tool for analyzing arguments. By assigning propositional variables to the statements making up the argument and by relating these variables by the appropriate operators, a generalized form of the particular argument is arrived at which can be tested to see whether it is a tautology, a contingency, or a self-contradiction. The tautologous case is the most pertinent one, for this means that any argument like the one so formulated must be valid. A self-contradiction reveals that such an argument can never be correct. A contingency signifies that under some conditions, *i.e.*, under some truth assignment of the constituent variables, the conclusion is true or at least one of the premises is false, but that under other assignments, the premises are true and the conclusion false; this last is another way of saying that the argument is invalid.

[1]The compound statements studied in the first chapter, since they are *single* statements, can of course be represented by a single variable such as $P$.

The language of symbolism, then, and the truth table analysis of its statements can be used immediately to determine whether specific arguments are valid. An illustration:

> To say that Washington successfully cornered Cornwallis is to say that he outgeneraled him.
>
> Had Cornwallis not been so cornered, there would have been no surrender at Yorktown.
>
> Hence, as surely as there was a Yorktown surrender, Washington outgeneraled Cornwallis.

The argument is symbolized thus:

$$p \equiv q$$
$$\bar{p} \supset \bar{r}$$
$$\therefore r \supset q$$

which can be written as one statement: $p \equiv q \cdot \bar{p} \supset \bar{r} . \supset r \supset q$. All that is lacking is to test this statement to see if it is a tautology.

### EXERCISE II–1

1. Use the truth table as suggested above to test the argument about Washington.
2. Test similarly these arguments:
   a. Since French perfumes enjoy such prestige, Joyce can be expected to buy them if she uses perfume. She buys no French perfumes. Therefore, she must not use perfume. (A review of question 1-n on page 16 may be of help.)
   b. Joyce is a good girl and she's four years old and she doesn't use perfume. Hence, she's either a good girl or her father wishes she would grow up.
   c. If Annette is younger than Harry, you may be sure he has fallen in love with her. He has indeed fallen in love with her. So she must be younger than he.

## 2. The Trapezoid Technique

In the previous section the truth table was used to determine whether a statement is tautologous, contingent, or self-contradictory. There are easier-to-use alternative devices which should be mastered without more delay. The first of these rewrites the vertical column from any part of a truth table in a relatively compact diagram based on an inverted trapezoid.[2] If each side of this diagram be regarded as corresponding to a line of the truth table for two variables thus:

4
3 2
1

| | | $q$ | $p$ |
|---|---|---|---|
| the bottom side | line 1 | 1 | 1 |
| the right side | line 2 | 1 | 0 |
| the left side | line 3 | 0 | 1 |
| the top side | line 4 | 0 | 0 |

it becomes an easy matter to represent any column in the truth table simply by writing the sides of the trapezoid corresponding to the lines marked *1* and omitting the sides corresponding to lines marked *0*.

By this convention $p$ can be represented by writing the first and third sides, *i.e.*, the sides corresponding to the lines of the table in which $p$ is *1*. By writing the first and second sides, $q$ is represented.

$p$: $q$:

Consider next how the negations of these variables must be written. Clearly the trapezoid symbols will be the complements of those below, *i.e.*, will contain those lines which before were missing and lack those lines which were present.

$\bar{p}$: $\bar{q}$:

Were the expression $p \vee q$ to be written using these symbols of letters to represent the variables, the result would be

$p \vee q$: $\vee$

Constituting as it does a substitute for the lines of the truth table, the trapezoid can be put to the uses of the table provided there is some way of writing a new symbol in the place of two previous ones. To take a very

[2]The inverted trapezoid was conceived by W. T. Parry ("A New Symbolism for the Propositional Calculus," *Journal of Symbolic Logic*, Vol. 19, #3, 1954, 161–168) as a device for representing all the 16 possible relationships between $p$ and $q$, *i.e.*, all the truth table analyses that can be written using *1* and *0* in their various combinations on four lines. These are:

the 1 combination in which all 4 lines are *1*,
the 4 combinations in which only 3 lines are *1*,
the 6 combinations in which only 2 lines are *1*,
the 4 combinations in which only 1 line is *1*, and
the 1 combination in which no line is *1*.

Of these we have put to use only four relations: $p \vee q$ (represented on Parry's trapezoid as $p$ $q$), $p \cdot q$ ($p$ $q$), $p \supset q$ ($p$ $q$), and $p \equiv q$ ($p$ $q$). Although the remaining 12 can each be expressed by means of these operators and negation (or even more simple example, once the table is set up with columns giving the values of $p$ and of $q$, it is possible to write in another column showing the value of $p \vee q$.

| $q$ | $p$ | $p \vee q$ |
|---|---|---|
| 1 | 1 | 1 |
| 1 | 0 | 1 |
| 0 | 1 | 1 |
| 0 | 0 | 0 |

Is it possible to perform the corresponding operation with trapezoid symbols, *i.e.*, given _ as an equivalent of the middle column in the table and _/ as the equivalent of the left column, can some symbol corresponding to the column under the wedge be derived from these? For if this merging of two symbols to form a new one representing their disjunction is possible, and like mergings are possible for representing the conjunction of two symbols, their equivalence, and implication, then the trapezoids can become a most useful instrument for performing truth table operations.

How to merge _ and _/ into a new symbol representing their disjunction can be deduced from the truth table procedure in filling in the column under the wedge. There we enter a *1* provided either of the original columns shows a *1*. This suggests that the new symbol we are seeking will have a side wherever either of the two component symbols has a side. In short, the rule for so merging two trapezoid symbols over a wedge is simply to superimpose the two symbols, for this will produce exactly the symbol desired.

$p \vee q$: _ ∨ _/ becomes _/

$p \vee \bar{q}$: _ ∨ \‾ becomes \‾_

$\bar{p} \vee \bar{q}$: ‾/ ∨ \‾ becomes \‾/

The same kind of merging of the component symbols when they are related by the horseshoe is slightly more complicated but still easy. The first tautology appearing in the first chapter

$$\bar{p} \vee q \,.\!\equiv p \supset q$$

suggests the rule for merging over implication: write the complement of the antecedent and superimpose the consequent. There is no need to write in the

---

economically), it is nevertheless systematic and elegant for a single matrix (the trapezoid symbol) to provide every operator possible, even though some are ordinarily nameless. The symbol is ingeniously designed to resemble the principal operators in their conventional form (see above).

The purpose to which the same symbol is put here (that of analyzing statements) uses the diagram as a substitute for the *variables* instead of for the *operators*. The present technique first appeared in the author's "Ideographic Computation in the Propositional Logic" in the same volume of the periodical cited above, pp. 169–171.

expression $\bar{p} \vee q$ as an intermediate step. The complementing of the antecedent expression is easily done mentally.

$p \supset q$ is written which merges to

$q \supset p$ is written which merges to

$\bar{q} \supset \bar{p}$ is written which merges to

In filling in the truth table under a dot, only such lines as are marked *1* for both components come to be marked *1* for the new statement. Accordingly, for merging over a dot, no side should be reproduced unless it appears in each component part, *i.e.*, the resultant figure should show only the *common* lines.

$pq$: becomes [3]

$\bar{p}q$: becomes

$\bar{q}\bar{p}$: becomes

Finally, since in filling in a truth table under an equivalence a *1* is entered in each line in which the two components are alike, whether they both be true or both be false, the resultant trapezoid symbol will contain a side for each common side (those cases in which each component is true) and also for each side commonly missing (those cases in which each component is false).

$p \equiv \bar{q}$: becomes

$\bar{p} \equiv q$: becomes

$\bar{p} \equiv \bar{q}$: becomes

Before this technique can be of much practical use it must be extended to more than two variables. This is done by expanding the symbolism in the same way the truth table it represents is expanded. It will be remembered that the form of the truth table used in Chapter I is such as to allow for expansion without any modification of the original table. The full benefits of this are now to be reaped. For what is needed here is an extension of the trapezoid symbol which will leave intact the now familiar symbols for $p$ and for $q$. These symbols must naturally be larger — they are each now true on four lines (of the eight) instead of only two lines (of the four), but their form is the same.

[3]The student need exercise only reasonable care in writing these symbols. If the slant of the side lines is preserved there will be no mistaking the left side for the right. Similarly, the length of the line, as well as position, helps to distinguish the top from the bottom.

| $r$ | $q$ | $p$ |
|---|---|---|
| 1 | 1 | 1 |
| 1 | 1 | 0 |
| 1 | 0 | 1 |
| 1 | 0 | 0 |
| 0 | 1 | 1 |
| 0 | 1 | 0 |
| 0 | 0 | 1 |
| 0 | 0 | 0 |

4 8
3 2 7 6
1 5

$p$:

$q$:

$\bar{p}q$:

The symbol required for *r*, which is true in the first four lines of the table and false on the next four, requires some convention for indicating the place occupied by a sideless trapezoid. The *x* serves for this.

$r$: *x* $\bar{r}$: *x*

For four variables, the truth table requires sixteen lines and the trapezoid symbol sixteen sides. Again, each previous symbol is simply expanded by doubling. The symbol for *s* is obvious.

$p$:

$q$:

$r$: *x* *x*

$s$: *x* *x*

Subsequent adaptations to larger numbers of variables follow like patterns of doubling for each symbol. By using circles (they are more swiftly written) for complete trapezoids, dashes to indicate repetitions of a symbol or of an obvious pattern of symbols, vertical lines to mark off the trapezoids into groups of eight, and other such devices as practice suggests,[4] it is not impractical to work with as many as thirty-two trapezoids. Thus *ru*, when seven variables are involved, is written quickly:

○ *x* ○ *x* —— | *x* *x* —— | ○ *x* ○ *x* —— | *x* *x* ——

[4]The conjunction of several premises is often facilitated by writing their (final) symbols under each other. The conjunction can then be written out at the bottom (as in addition) by setting down the lines common to all.

1st premise: *x*

2nd premise:

3rd premise: *x* ○ ○ ○

Conjunction: *x* *x*

The following illustrates the successive mergings by which a complex expression is reduced to a single symbol.

> One leaves the coral snake alone, or someone gets bitten.
>
> If someone gets bitten, either the physician or the undertaker gets some business.
>
> ∴ If one molests a coral snakeand the undertaker gets no business, the physician does.

The argument as a single statement:

$$p \vee q : q \supset . r \vee s :\supset \bar{p}\bar{s} \supset r$$

The same statement expressed in trapezoids:

$p \quad \vee \quad q \quad : \quad q \quad \supset . \quad r \quad \vee \quad s \quad :\supset$

∖_ ∖_ ∖_ ∖_ ∨ _/ _/ _/ _/ : _/ _/ _/ _/ ⊃. O x O x ∨ O O x x :⊃

$\bar{p} \quad \cdot \quad \bar{s} \quad .\supset \quad r$

‾/ ‾/ ‾/ ‾/ · x x O O .⊃ O x O x

Notice that the trapezoid expression has a dot between $\bar{p}$ and $\bar{s}$. This is not indispensable, but it does prevent the juxtaposition of the two symbols which might thereby resemble a single symbol for five variables. This dot between $\bar{p}$ and $\bar{s}$ calls for a dot beside the following horseshoe so that the latter may keep its higher rank. Punctuation is obviously just as important when working with trapezoids as it ever is. After merging over the lesser operators, this is what results:

∖_/ ∖_/ ∖_/ ∖_/ : _/ _/ _/ _/ ⊃. O O O x :⊃ x x ‾/ ‾/ ⊃ O x O x

Further mergings produce still shorter expressions until only one symbol is left. The previous writing merges to

∖_/ ∖_/ ∖_/ ∖_/ : O O O ∖‾ :⊃ O O O ∖_

which in turn becomes

∖_/ ∖_/ ∖_/ ∖ :⊃ O O O ∖_

which when finally merged over the major horseshoe is

The final symbol indicates that the expression in question (the argument) is true on every line of the truth table, which is to say that it is tautologous, or valid.

### Rules for Manipulating Trapezoid Symbols

| | |
|---|---|
| Negation: | Complement the symbol |
| Disjunction: | Superimpose |
| Implication: | Complement antecedent; superimpose consequent |
| Conjunction: | Common lines only |
| Equivalence: | Common lines and lines commonly missing |

## EXERCISE II–2

Using trapezoids, show that each of the following statements is a tautology:

1. $p \supset q .\equiv \bar{p} \vee q$
2. $p \equiv q .\equiv pq \vee \bar{p}\bar{q}$
3. $\sim(pq) \equiv . \bar{p} \vee \bar{q}$
4. $\sim(p \vee q) \equiv \bar{p}\bar{q}$
5. $p \cdot q \vee r .\equiv pq \vee pr$
6. $p \vee q \cdot p \vee r .\equiv p \vee qr$
7. $\sim(p \supset q) \equiv p\bar{q}$
8. $\sim(p \equiv q) \equiv . p\bar{q} \vee \bar{p}q$

Determine whether each of the following is a tautology, a contingency, or a self-contradiction.

9. $p \vee q .\supset p$
10. $p \vee . q \supset p$

11–14. Test each of the arguments in the exercise on page 21, this time using trapezoids.

*15. Sid thought the proper way of negating $pq$ was to write $\bar{p}\bar{q}$. Use trapezoids to determine whether he was right.

### 3. Transformations

Truth tables (and the variation called the trapezoid method) can determine whether a given expression is a tautology, contingency, or self-contradiction. There is another procedure which can handily identify self-contradictions. Because this method consists of transforming the expression according to rule in much the same way as an algebraic equation is transformed by transposition, factoring, and the like, it will constitute an introduction to the *manipulation* of the symbolism introduced in Chapter I. Such manipulation underlies the line-by-line deductive proofs to be undertaken later.

For the present the purpose of these transformations will be to reduce a given expression to what is called a *disjunctive normal form* (DNF), which consists of nothing more than literals and wedges. Thus each of the following is in disjunctive normal form:

$$p$$

$$\bar{p}r$$

$$pq \vee prs \vee \bar{q}\bar{r}$$

$$r \vee pqs \vee q\bar{s}t \vee q\bar{s}\bar{u}$$

Another definition of the DNF is that it consists of one or more *disjuncts*, each consisting of either a literal or a conjunction of literals. Strictly speaking, a disjunct is an expression immediately under a wedge; as a wedge must relate two such component expressions, it is irregular to speak of *one* or more disjuncts. But it will be very convenient to allow ourselves the slight impropriety of sometimes considering an expression standing under no operator as a 'wedgeless disjunct.' Thus $p$ or $\bar{p}r$ above can be so considered. The fourth DNF above has four regular (*i.e.*, wedged) disjuncts. A *conjunct* is an expression immediately under a conjunction. So a DNF can be defined by the order of its operators, or by listing the operators themselves, since an unwritten conjunction, the bar, and the wedge can only order themselves in the proper way.

Since this is what a DNF is, it is not surprising that any self-contradictory expression put into this form must bear its character on its sleeve. A disjunction asserts that one or another of its disjuncts is true; hence for the DNF to be self-contradictory, *each* of the conjunctions comprising its disjuncts must be self-contradictory. And the conjunction comprising the disjunct is self-contradictory if and only if it contains two contradictory literals, whatever else it may contain. The DNF $p\bar{p} \vee \bar{q}qrs$ is self-contradictory only because each of its disjuncts is.

The process of converting a given expression into DNF consists of ridding it of horseshoes, equivalences, curls, and conjunctions having dots. Such eliminations must leave only wedges, unwritten conjunctions, and bars as the operators; these, as stated above, necessarily order themselves in the proper way. This process of conversion can be divided into three steps. The first will remove any horseshoes or equivalences, together with any curls that might stand immediately over them. The second will remove any curls standing over conjunctions or disjunctions. The last will convert any dot-over-wedge expressions into wedge-over-conjunction ones. By these eliminations of unwanted operators, the DNF must emerge. Explicitly, the steps are:

1. Rewrite the statement, substituting for each implication or equivalence the appropriate equivalent from these patterns:[5]

| | |
|---|---|
| 1a. $p \supset q \mathbin{.}\!\equiv \bar{p} \vee q$ | TH (Transformation of horseshoe) |
| 1b. $\sim(p \supset q) \equiv p\bar{q}$ | TH |
| 1c. $p \equiv q \mathbin{.}\!\equiv pq \vee \bar{p}\bar{q}$ | TE |
| 1d. $\sim(p \equiv q) \equiv\!\mathbin{.}\; p\bar{q} \vee \bar{p}q$ | TE |

This first step may require more than one rewriting to complete, as when, for example, equivalences or horseshoes occur under equivalences or horseshoes. Once the step is finished, the expression contains only dots, wedges, and negations.

2. Rid such an expression of curls by using what are called De Morgan transformations (after the logician whose theorems first adduced them):

2a. $\sim(pq) \equiv\!\mathbin{.}\; \bar{p} \vee \bar{q}$ 2b. $\sim(p \vee q) \equiv \bar{p}\bar{q}$ TM

These paradigms express a principle which can be applied readily to quite complex expressions. The negation of any statement comprised solely of dots, wedges, and negations is an expression in which each wedge not subordinate to a curl has been changed to a dot, each dot not subordinate to a curl has been changed to a wedge, and each variable (or parenthesis under a curl) changes its sign. Accordingly,

$$\sim(p\bar{q} \vee \mathbin{.}\, r \cdot \sim(s \vee \bar{t}u)) \qquad \text{is equivalent to} \qquad \bar{p} \vee q \cdot \bar{r} \vee s \vee \bar{t}u$$

Notice that the rank of each operator is given to the operator replacing it.

Here again, should there be parentheses within parentheses, more than one transformation may be required to complete this step. When finished, the only negations remaining are bars; the only feature preventing the expression from being a DNF (if it is not one already) is the subordination of some of the wedges to dots.

The third step repairs this deficiency by *distributing*. In algebra an expression such as $3y + x$ can be multiplied by $x$ to yield $3xy + x^2$ or by

[5] The word *pattern*, for which the synonym *schema* (pl.: *schemata*) will generally be substituted, introduces tacitly a principle underlying all symbolic logic. For the occurrence of $p$ in such a pattern, any variable (including $p$ itself) or any statement can be substituted provided only that it is substituted throughout the pattern. Any variable or expression can similarly be substituted for $q$. Thus, in applying the schema $\sim(p \supset q) \mathbin{.}\!\equiv p\bar{q}$ to $\sim(r \vee \bar{p} \mathbin{.}\!\supset q)$ in order to rewrite it as $r \vee \bar{p} \cdot \bar{q}$, the $p$ of the schema is allowed to stand for the expression $r \vee \bar{p}$, while $q$ stands for $q$.

This principle of substitution is simply an extension of the use of variables to represent simple propositions. The variable is now allowed to stand for compound statements.

$x + 2$ to yield $3xy + 6y + x^2 + 2x$. In propositional logic a like multiplying-out procedure can weld a disjunction to a conjoined statement.

3. Rid the expression of dots over wedges by distributing according to this pattern:

$$\text{3a. } p \cdot q \vee r .\equiv pq \vee pr \qquad \text{TD}$$

Another TD schema, which will become useful in the ensuing sections, is

$$\text{3b. } p \vee qr \equiv . p \vee q \cdot p \vee r \qquad \text{TD}$$

Schema 3a subtends such instances as the transformations of

$$p \vee q \cdot r \vee s \quad \text{into} \quad pr \vee qr \vee ps \vee qs$$

or of

$$p \vee q \cdot r \vee s \vee tu \quad \text{into} \quad pr \vee qr \vee ps \vee qs \vee ptu \vee qtu$$

Each distribution of a disjunction over that to which it is conjoined eliminates that conjunction; the number of distributions required thus depends on the number of conjunctions standing over wedges in the expression at the end of the previous step.

Since any conjunction in which the same letter occurs both with and without a bar is self-contradictory, such a conjunction occurring in this final step is simply deleted, unless the entire expression is made up of such conjunctions; in such a case one such expression is preserved. By this procedure the DNF of a self-contradictory expression will emerge as $p\bar{p}$ or as $\bar{q}rqs$ or some such single conjunction standing as a wedgeless disjunct. To illustrate the sort of deletion thus justified, observe that $p \vee qs \cdot \bar{p} \vee \bar{s}$ becomes $p\bar{p} \vee qs\bar{p} \vee p\bar{s} \vee qs\bar{s}$ when distributed, but as the first and last disjuncts of this DNF are self-contradictory, a simpler writing would be $qs\bar{p} \vee p\bar{s}$.

Similarly, a DNF containing two contradictory disjuncts, each of a single literal, is plainly tautologous. All *other* disjuncts may then be omitted without affecting the statement's truth value. Thus $p \vee : \bar{s} \supset . qr \vee \bar{s}$ is transformed by TH to $p \vee : s \vee . qr \vee \bar{s}$ in which the dots are no longer needed and the nature of the second and last disjuncts allows the others to be discarded so that the writing $s \vee \bar{s}$ is adequate.

Furthermore, whenever a literal is conjoined to itself, its second occurrence need not be set down, since $p$ is the equivalent of $pp$. The expression $pqr \equiv pqs$ becomes (by TE) $pqrpqs \vee \sim(pqr)\sim(pqs)$, which is more simply written as $pqrs \vee \sim(pqr)\sim(pqs)$.

Finally, if two disjuncts are exactly alike, only one need be written. For $pq \vee r \vee pq$, one may as well write only $pq \vee r$.

Below is an illustrative reduction of a statement to DNF. By carefully following the details of each step, the reader will understand the applications of the schemata more fully. Given:

$$\text{1. } \quad p \cdot q \equiv \bar{r}t \supset . p \supset q\bar{s} \cdot \bar{p} \vee \sim(\bar{r}t)$$

Recourse to TH (pattern 1a) yields:

$$2.\quad \sim(p \cdot q \equiv \bar{r}t) \vee . p \supset q\bar{s} \cdot \bar{p} \vee \sim(\bar{r}t)$$

By TE (1c) and TH:

$$3.\quad \sim(p \cdot q\bar{r}t \vee \bar{q}\sim(\bar{r}t)) \vee . \bar{p} \vee q\bar{s} \cdot \bar{p} \vee \sim(\bar{r}t)$$

which completes Step 1. By using TM:

$$4.\quad \bar{p} \vee . \bar{q} \vee r \vee \bar{t} \cdot q \vee \bar{r}t \vee : \bar{p} \vee q\bar{s} \cdot \bar{p} \vee r \vee \bar{t}$$

This completes Step 2. TD then yields:

$$5.\quad \bar{p} \vee . rq \vee \bar{t}q \vee \bar{q}\bar{r}t \vee : \bar{p} \vee q\bar{s}\bar{p} \vee \bar{p}r \vee q\bar{s}r \vee \bar{p}\bar{t} \vee q\bar{s}\bar{t}$$

completing Step 3 and providing the DNF, although in a form still open to improvement, not only by eliminating the now redundant punctuation dots, but also by alphabetizing the letters within each conjunction. This last does more than satisfy the sense of elegance; it makes further simplification easier.

### 4. The Simplest Disjunctive Normal Form

The four deletions sanctioned in the previous section can be regarded as rudimentary ways of *simplifying* a DNF — so rudimentary, indeed, that schemata were not even offered. This omission of schemata merely means that such transformations by simplification (TS) need not be separately written out. If one chooses to write such simplifications on successive lines, however, such lines can quite properly be construed as appeals to TS. Thus if

$$prs \vee qq \vee \bar{r}rs \vee prs \vee p$$

has been written, this subsequent line is altogether regular:

By TS $\quad prs \vee q \vee p$

Or from

$$pq \vee \bar{q} \vee \bar{p}r\bar{s} \vee q \vee q\bar{r}$$

one can go to:

By TS $\quad \bar{q} \vee q$

But there are further, rather less elementary, ways of eliminating redundancies in a DNF. For these varieties of TS we had better use schemata. The first of three is

4a. $pq \vee p . \equiv p$ TS

This sanctions the deletion of any disjunct a component part of which occurs disjoined. By construing *a component part* as including the whole, the rudimentary reduction of $pq \vee r \vee pq$ to $pq \vee r$ (mentioned above) becomes a special case under this schema. Simplifications like the following, however, are more typical:

$$pqr \vee pq \quad \text{simplifies to} \quad pq$$
$$q\bar{r}u \vee \bar{t} \vee pq\bar{r}su \quad \text{simplifies to} \quad q\bar{r}u \vee \bar{t}$$

In the next schema, it is not a disjunct which is deleted, but a literal within a disjunct, provided that literal is negated in another disjunct containing — apart from that negation — no literals not found in the first.

$$\text{4b. } \bar{p}q \vee p \mathrel{.}\equiv q \vee p \qquad \text{TS}$$

This pattern can be better understood if these applications are examined:

$$pqs \vee \bar{p} \quad \text{simplifies to} \quad qs \vee \bar{p}$$
$$pqr\bar{s} \vee pqrs\bar{t}\bar{u} \quad \text{simplifies to} \quad pqr\bar{s} \vee pqr\bar{t}\bar{u}$$

In practice the rule amounts to this: *one* literal can be struck from the *longer* of two disjuncts (the disjuncts can be of the same length, but this occurs less often) provided that literal occurs negated in the shorter and the other literals of the shorter are each to be found in the longer. Between them, this schema and 4a simplify $pq \vee p\bar{q}$ to $p$, the intermediate step being $pq \vee p$.

These two schemata are sufficient for simplifying the ponderous expression (5) at the end of the previous section to

$$\text{6. } \bar{p} \vee qr \vee q\bar{t} \vee \bar{q}\bar{r}t$$

But they are powerless[6] to simplify

$$\bar{p}r \vee p\bar{q} \vee q\bar{r} \vee p\bar{r} \vee \bar{q}r \tag{1}$$

which has as a simple form the first three disjuncts only. This simplification can be effected by some careful scrutiny and recourse to this third schema:

$$\text{4c. } pq \vee \bar{p}r \vee qr \mathrel{.}\equiv pq \vee \bar{p}r \qquad \text{TS}$$

by which is justified the deletion of any disjunct composed of conjuncts elsewhere conjoined to contradictory literals. The expression (1) can be rid of its fourth disjunct, $p\bar{r}$, by the presence of the second, in which the $p$ of this conjunction is found conjoined to $\bar{q}$, and the third, in which $\bar{r}$ is found conjoined to $q$. Similarly the first and second disjuncts eliminate the last. By this schema

$$pqs \vee \bar{p}rs \vee qrs \quad \text{simplifies to} \quad pqs \vee \bar{p}rs$$

by regarding $qrs$ as $qsrs$, to which it is of course equivalent. The expression $pqr \vee \bar{p}\bar{q}s \vee rs$ does *not* qualify for simplification by this schema because

[6]Later in this section it will be shown that there are ways of effecting any simplification with only these two schemata (*cf.* note 9 on page 36); but the uses so far assigned them fall short of the task.

the expressions to which the conjuncts $r$ and $s$ are elsewhere conjoined are not contradictory.

By adding this schema and using it in this way to eliminate disjuncts (another use will be discussed later) the range of expressions that can be simplified has been widened. But with this schema the need for insight has been introduced if the shortest form of a DNF is to be reached. Schemata 4a and 4b can be applied on any impulse, so to speak, with the same effect. But if in attacking (1) with schema 4c, the fourth and fifth disjuncts are used to eliminate the second, the expression becomes incapable of further reduction (apart from devices yet to come) despite the fact that it is not yet in simple form.

This contrast between procedures that require no guiding insight and those that do is of much concern to logicians. The procedure which requires no insight, but is instead entirely mechanical, is called an *effective* one, the word here having a technical meaning. Thus an arithmetical process such as addition or multiplication is effective. One uses his head, to be sure, but not in the same way he must use his head in writing a poem or sculpting a figure. The procedure which is effective can be turned over to a machine having no creative ability but capable of following instructions. There is a close connection between an effective procedure and what the mathematician calls an algorithm, a step-by-step procedure which is never ambiguous as to what is to be done next and succeeds in accomplishing some specified task. Thus reducing any expression to a DNF is an effective procedure: by following the steps given in the previous section, one can always bring about the transformation. But the process of reducing a DNF to a simple form — one having as few disjuncts as any other equivalent DNF — cannot be made effective without going into the matter still farther.

The only way so far discovered of knowing that a particular DNF is simple is to compare it with other equivalent DNFs, the given one being simple if no other is simpler. But to make this comparison, the alternative DNFs must be on display in some way. What this means can be seen by examining the trapezoid symbol for (1).

Recalling that each side of the trapezoid symbol corresponds to a line of the truth table, let the particular truth values of each variable be set beside the sides of the symbol.

$$\bar{p}\bar{q}r$$

$$p\bar{q}r \quad \bar{p}qr \quad p\bar{q}\bar{r} \quad \bar{p}q\bar{r}$$

$$pq\bar{r}$$

The expressions so set down are known collectively as the Boolean expansion of (1), since each represents but a single line on the truth table. The Boolean expansion is the first tool to be used to put on display the equivalent DNFs that are needed.

It follows from the rule for merging trapezoid symbols over the horseshoe that one expression, *A*, implies another, *B*, if the trapezoid symbol for *A* is included in that for *B*. By this it is apparent that each of the expressions in the Boolean expansion must imply (1). But there are expressions of fewer literals which are also implicants[7] of (1). For instance $p\bar{q}r$ and $\bar{p}\bar{q}r$ can be embraced in one expression (*cf.* the trapezoid symbol): $\bar{q}r$. If a still more abbreviated expression, *e.g.*, $\bar{q}$, or $r$, be considered, it will be noticed that it is not an implicant of (1), *i.e.*, neither the trapezoid symbol for $\bar{q}$ nor that for $r$ is included in the symbol for (1). Since $\bar{q}r$ cannot be shortened, it is called a *prime implicant* of (1). These prime implicants are essential to the process of determining whether a DNF is simple because they are the expressions which must be used if a DNF is to be economical. A simple DNF will use just enough of these prime implicants to express what has to be expressed. *If all the prime implicants are listed, then a simple DNF will be a most economical selection from that list.* This list, therefore, with the alternative choices it makes possible, is the mode of displaying all the alternative DNFs that are equivalent.

How to obtain all these prime implicants, then, becomes an important part of the procedure for arriving at a simple DNF. They are got by using schema 4c in reverse, so to speak, not to eliminate disjuncts but to generate further ones, even though they may be redundant. If one carries out this process as far as possible, the while using schemata 4a and 4b whenever possible, the DNF will come to contain every one of its prime implicants. All of this will now be illustrated.

The first two disjuncts in (1) can be made to fit into the right side of schema 4c because they have exactly one literal in opposition.

$$\bar{p}r \vee p\bar{q} \vee \bar{q}r \mathrel{.{\equiv}} \bar{p}r \vee p\bar{q}$$

From the left side it will be seen that $\bar{q}r$ is the disjunct they generate. Since this disjunct is already present in the DNF, it need not be added. The first and third disjuncts generate $\bar{p}q$ in the same way. This is added:

$$\bar{p}r \vee p\bar{q} \vee q\bar{r} \vee p\bar{r} \vee \bar{q}r \vee \bar{p}q$$

The first and fourth disjuncts do not fit into the right side of the schema since they have two contradictory literals. The first and fifth are useless because they have no contradictory literals. The procedure requires that every pair contradictory in exactly one letter be tested to see if they yield a new disjunct. In our example, the process yields only the six disjuncts appearing above.

The next task is to find the smallest group of prime implicants equivalent to the whole. This is conveniently done by using a table with the expressions of the Boolean expansion across the top. These expressions must be 'covered' by the simple DNF. At the right, in the inner column, are listed the prime implicants, the expressions capable of doing the covering. The check marks

[7] *Implicant*, a synonym of *implicans*, has a somewhat more convenient plural form.

indicate which Boolean expressions are so covered by these prime implicants. Each of the latter will be seen to express whatever Boolean expressions contain its literals. Recourse to the trapezoid symbol will show why this is so.

| $pq̄r$ ∨ | $\bar{p}\bar{q}r$ ∨ | $\bar{p}qr$ ∨ | $p\bar{q}\bar{r}$ ∨ | $pq\bar{r}$ ∨ | $\bar{p}q\bar{r}$ | | |
|---|---|---|---|---|---|---|---|
| | ✓ | ✓ | | | | $\bar{p}r$ | $p'$ |
| ✓ | | | ✓ | | | $p\bar{q}$ | $q'$ |
| | | | | ✓ | ✓ | $q\bar{r}$ | $r'$ |
| | | | ✓ | ✓ | | $p\bar{r}$ | $s'$ |
| ✓ | ✓ | | | | | $\bar{q}r$ | $t'$ |
| | | ✓ | | | ✓ | $\bar{p}q$ | $u'$ |

Inspection of this table reveals that the first three prime implicants subtend the entire Boolean expansion, and that the last three do also. For (1), then, two simple forms exist: $\bar{p}r \vee p\bar{q} \vee q\bar{r}$ and $p\bar{r} \vee \bar{q}r \vee \bar{p}q$.

Even this inspection can be made mechanical. By referring to the prime implicants by means of the primed variables written to their right, let a disjunction be made of the prime implicants checked in each column, and let these disjunctions be conjoined:

$$q' \vee t' \cdot p' \vee t' \cdot p' \vee u' \cdot q' \vee s' \cdot r' \vee s' \cdot r' \vee u'$$

This expression, by the way, is in *conjunctive* normal form (CNF), *i.e.*, a conjunction of disjunctions, each disjunct of which consists of a single literal. By TD this expression is then converted into DNF (which need not be reduced to simplest form):

$$p'q'r' \vee p'q'r't' \vee q'r't'u' \vee p'q'r's' \vee p'r's't' \vee$$
$$r's't'u' \vee p'q's'u' \vee p's't'u' \vee s't'u'$$

The smallest groups of prime implicants are here indicated by the *shortest disjunct(s)*, namely, the first and the last. Should two or more equally short disjuncts indicate groups of prime implicants differing in the total number of literals, the group containing fewest literals can be chosen.[8]

Because each of the steps is a mechanical one, this method for simplifying a DNF is effective. The procedure can be set down in four steps: (1) Write the Boolean expansion of the DNF to be simplified. (2) List all its prime implicants (designating each by a variable). (3) For each Boolean expression make a disjunct of each prime implicant that it implies, using the above designations for the prime implicants. Conjoin these disjunctions. (4) By TD convert this into a DNF. The shortest disjunct(s) in this DNF indicates which prime implicants constitute a simple DNF of the original expression.

[8]In the procedure just explained, the tabulation of disjuncts and prime implicants is due to W. V. Quine ("The Problem of Simplifying Truth Functions," *American Mathematical Monthly*, Vol. 59, 1952, 521–531) and the device for mechanically selecting the shortest list of prime implicants is due to E. J. McCluskey, Jr. ("Minimization of Boolean Functions," *Bell System Technical Journal*, Vol. 35, 1956, 1417–1444).

Inasmuch as the generative use of schema 4c, together with 4a and 4b, is sufficient for producing all the prime implicants (the critical step in this procedure), it is plain that these three schemata are in this way adequate for the task of simplification.[9] But although this method is an effective one, in practice it is common to pursue swifter ones when possible.[10]

The Veitch charts and Karnaugh (pronounced *Kar-no*) maps designed for the use of engineers are two such practical methods for simplifying DNFs. The trapezoids can also be exploited as such a method. A tautology, once cast into trapezoid symbols, is patent; the same is true of any self-contradiction. The question is whether a contingency expressed in trapezoids can be useful in writing a simple DNF. Because it can, one advantage appears immediately, namely, that in seeking a simple DNF by first expressing a statement in trapezoids no intermediate reduction to DNF is required. A complex expression involving operators other than wedges and unwritten conjunctions can be reduced by the usual mergings over all operators to a trapezoid symbol, that symbol can then be used to arrive at the prime implicants (also expressed in trapezoids), and these symbols can then be used to discern a simple DNF. This procedure will now be explained. Although a high degree of competence in the use of the method will appeal primarily to the engineer, the logic student will at least be concerned with the principles of the technique.

[9]This was proved independently and at about the same time by Edward W. Samson and Burton E. Mills ("Circuit Minimization: Algebra and Algorithms for New Boolean Canonical Expressions," *AFCRC Technical Report 54–21*) and by Quine ("A Way to Simplify Truth Functions," *American Mathematical Monthly*, Vol. 62, 1955, 627–631). In Quine's terminology the $rs$ generated by $pr$ and $\bar{p}s$ is called their *consensus*. (The term *prime implicant* was also introduced into common use by Quine.)

Strictly speaking, because the schema 4b can be derived from 4c and 4a, and 4c can be derived from 4b (used generatively) and 4a, 4a coupled with either of the others is adequate for the task of simplification. Thus 4a and 4b are sufficient (with insight) to simplify the expression (1) on page 32:

$$1.\ \bar{p}r \vee p\bar{q} \vee q\bar{r} \vee p\bar{r} \vee \bar{q}r$$

The second disjunct can be used to modify the fourth; and the first, the last:

$$\text{By 4b } 2.\ \bar{p}r \vee p\bar{q} \vee q\bar{r} \vee pq\bar{r} \vee p\bar{q}r$$

whereupon the last is deletable by the second and the fourth by the third:

$$\text{By 4a } 3.\ \bar{p}r \vee p\bar{q} \vee q\bar{r}$$

Even the complete list of prime implicants can be generated with 4a and 4b or with 4a and 4c.

[10]In 1938 there appeared in the *Transactions of the American Institute of Electrical Engineers* (Vol. 57, 713–723) a landmark article by Claude E. Shannon, "A Symbolic Analysis of Relay and Switching Circuits," which provided modern logic with one of its rare direct applications to technology. Shannon showed that any relay circuit has a corresponding expression in the propositional logic. It follows that once a circuit's equivalent logical expression is set down, any simplification of the expression signifies a possible simplification of the circuit.

Since that time an entire new industry has burgeoned around digital computers, in whose early evolution various logic machines figured importantly and in whose design and use logic plays a principal role. (*Logic Machines and Diagrams*, Martin Gardner, New York: McGraw-Hill, 1958, is an easily read account of this role.) The practical need for simplifying logical expressions not only gave added importance to the work of Quine and other theoreticians, but brought into being the diagrams and maps referred to. These exist to facilitate simplifications. A perusal of *Switching Circuits for Engineers*, Mitchell P. Marcus (Prentice-Hall, 1967) will reveal the importance of such simplifications. Circuit design contemplates several varieties of simple forms, the DNF and CNF being only two of them.

It was remarked earlier that any expression symbolized in trapezoids has as an implicant any other expression whose trapezoid symbol is *contained therein* and that consequently any expression has among its implicants each expression of its Boolean expansion. For example, among the implicants of

‾7 _/ (1)

are \ x as well as — x, and / x, and x \, and x _ , and x /. And these can be seen to be implicants of the given expression without ever going into the question of how either the implicant or the given expression can be put into literals. But of course any of these implicants can easily be so expressed. The single line constituting the symbol \ x can be said to be the $p$ (for observe that this line is part of the symbol for $p$: _ _ ) that is *in non-q* (since it is not within the symbol for $q$: _/ _/ ) that is *in r* ( _7 x ). And once the student is practiced in *mentally comparing a line in question with these patterns of the symbols for the variables*, he will have no difficulty seeing that (to give examples):

x — x x is the *non-p* in *non-q* in *non-r* in *s*: $\bar{p}\bar{q}\bar{r}s$

x _ is the *p* in *q* in *non-r*: $pq\bar{r}$

x x x x / x x x is the *non-p* in *q* in *r* in *s* in *non-t*: $\bar{p}qrs\bar{t}$

and so forth.

But this knowledge by itself, though it would enable him to write certain implicants (namely, those of the Boolean expansion) would not be likely to yield *prime* implicants. Converting such an implicant into a prime implicant entails at one time a reduction of the number of literals in the conjunction, as was noted, and a corresponding *doubling* of the number of lines in the trapezoid symbol. A test as to whether an implicant is prime, then, is the question: can its trapezoid symbol be doubled in size by the addition thereto of related lines also included in the given statement? A simple example is afforded again by the expression (1). Can the implicant $p\bar{q}r$ ( \ x ) be doubled by adding a related line also included in ‾7 _/? The answer clearly depends on what a related line is. A partially generalized answer is this:

1. An adjoining line is related. (When such a line is added, the variable $p$ or $q$, as the case may be, is deleted from the literals conjoined.)
2. a. If $r$ be among the literals in the conjunction, a parallel line of the same length one ($2^0$) trapezoid to the right is related; if $\bar{r}$, then such a line in the trapezoid to the left is related. (When such a line is added, the variable $r$ is deleted from the conjunction.)
   b. When $s$ is one of the literals, a parallel line of the same length two ($2^1$) trapezoids to the right (or if $\bar{s}$, then two

trapezoids to the left) is related (and this variable can be struck from the conjunction).

c. When $t$ (or *non-t*) is one of the literals, such a line four ($2^2$) trapezoids to the right (or to the left) is related (and this variable becomes deletable).

And so on for $u$, the distance being $2^3$ trapezoids; $v$, $2^4$ trapezoids; etc.

In the case of (1) the implicant $p\bar{q}r$ ( \ x ) can be doubled by the addition of an adjoining line, yielding ⟍‾ x and permitting the deletion of $p$ from the conjunction so that only $\bar{q}r$ remains; or alternatively, by the addition of a parallel line in the trapezoid to the right, yielding \ \ in which case the deletion of $r$ leaves $p\bar{q}$. Each of these additions lies within the given (1) so the resultant expressions are still implicants thereof. Here is another example: the implicant _ x x x of the statement

$$\text{(2)}$$

can be expanded into _ x x x , shortening the conjunction to $prs$, and also into _ _ x x , which results in $pqs$.

Once a symbol consists of two lines, subsequent expansions of this sort must result in four lines, then eight, sixteen, etc., each expansion occasioning the deletion of another variable. The above definitions of *related*, contemplating as they do the doubling of a single line, require some reinterpretation. For $n$ lines to be so expanded, each of the $n$ lines must be related to another line occurring in the given statement in the same way. For example, having now the implicant $\bar{q}r$ ( ⟍‾ x ) of statement (1), any licit doubling must afford each of these lines an adjoining one (the presence of either of the variables $p$ or $q$ in the conjunction suggests looking here) or else two more lines just like these but in the trapezoid to the right (the presence of $r$ dictates this). In neither case, however, can the expansion be effected, for in the first case, in which we seek _/ x , and in the second, in which we seek ⟍‾ ⟍‾ , there is involved a line which lies outside the given statement ( ⟍‾7 _/ ). Accordingly $\bar{q}r$ ( ⟍‾ x ) is a prime implicant of (1). An alternative initial expansion of $p\bar{q}r$ ( \ x ) yielded $p\bar{q}$ ( \ \ ) in the previous paragraph. A doubling of this symbol would entail either _ _ or ⟍‾ ⟍‾ , *i.e.*, would entail the addition of lines *related to each other.* (The symbols _ ⟍‾ and ⟍‾ _ fail to do this and consequently are not expressible as conjunctions of literals.) Can \ \ be doubled into either _ _ or ⟍‾ ⟍‾ without incurring lines lying outside (1)? Is $p\bar{q}$ a prime implicant?

One of the implicants of (2) developed from _ x x x was _ x x x , $prs$. Each of these three literals suggests a possible doubling of this implicant. If the $p$ is to be struck, the doubling must consist of adjoining lines, but since _/ x x x does not lie wholly within (2), this is not an implicant. The literal $r$ can be struck if in the trapezoid to the right (*cf.* 2–a in the definition of related lines) the same two lines can be found. As __ x x does lie wholly within (2), *ps* is a legitimate implicant. (We shall inquire presently whether this be a prime implicant.) The third way of doubling _ x x x is suggested by the occurrence of *s* in the conjunction *prs*. Accordingly these same lines should be sought in the second trapezoid to the right (*cf.* 2–b in the definition), but since _ x _ x involves a line not found in (2) this expansion is not possible.

Pursuing now the further expansion of the implicant *ps* ( __ x x ), we are led by the conjunct *p* to search for _/ _/ x x whereas the presence of *s* suggests ____. But since neither of these symbols lies entirely within (2), *ps* is a prime implicant of (2).

If each single line in (2) is in this way expanded into a largest symbol containable within (2), all of the prime implicants will be listed:

| *Boolean-expansion Implicants* | | *Prime Implicants* | |
|---|---|---|---|
| _ x x x<br>\ x x x<br>x _ x x<br>x \ x x | Any of these lines become | __ x x | *ps* |
| x ‾ x x | becomes | x \‾ x x | $\bar{q}\bar{r}s$ |
| x x \ x | becomes | \ x \ x | $p\bar{q}r$ |
| x x x / | is itself a prime: | x x x / | $\bar{p}q\bar{r}\bar{s}$ |

Once all of the prime implicants are set down, the selection of a simple DNF consists of selecting as small a group as will, when superimposed (*cf.* the merging rule of disjunction), reconstitute the original symbol (2). Since each of the primes in this case contains a line not found in any other prime, all of them are required: the simple DNF of (2) is $ps \vee \bar{q}\bar{r}s \vee p\bar{q}r \vee \bar{p}q\bar{r}\bar{s}$. Example (1) is rather different. As previously noted, the primes are:

| | | |
|---|---|---|
| \‾ x | $\bar{q}r$ | $t'$ |
| \ \ | $p\bar{q}$ | $q'$ |
| ‾/ x | $\bar{p}r$ | $p'$ |
| / / | $\bar{p}q$ | $u'$ |
| x _ | $p\bar{r}$ | $s'$ |
| x _/ | $q\bar{r}$ | $r'$ |

Two equally simple DNFs can be constituted. Either can be selected from such a list of trapezoid symbols simply by superimposing the selected symbols as one proceeds.

Of course, if a list of primes is extensive and a simple DNF entails a large number of disjuncts, the selection is not necessarily an easy matter even with the help of these symbols. But since each single line in the above listing corresponds to an expression of the Boolean expansion, it is altogether possible to apply here the recourse described earlier for making the selection mechanically. A comparison of the above listing with the table on page 35 will reveal that the only differences are in the order of the primes and in the fact that the expressions of the Boolean expansion are to be identified here by their situation in the trapezoid symbol. The different order of the primes has been compensated for by assigning to the primes here the same primed variables which represented them there. If one selects the trapezoid symbols containing the line $p\bar{q}r$ ( $\setminus$ $x$ ), the same disjunction with which the CNF on page 35 begins ($q' \vee t'$) will appear.

Systematic inspection of the trapezoid symbols, then, affords a practical and effective way of listing all the prime implicants of a statement. And theoretically the selection of a simple DNF by the mechanical means just referred to rounds out the technique. But because the number of prime implicants in a complex statement of several variables can be staggering — easily outnumbering the lines in its truth table — one might in practice prefer a fallible guesswork selection of a DNF he hopes is simple to an infallible computation that is so tedious. From the logician's point of view a major problem respecting simplification, still unsolved, is *whether some technique can be adduced which obviates the exhaustive listing of prime implicants and is nevertheless effective*. Such a technique would enable the practicing engineer to shorten almost any technique he uses.

It was stated at the beginning of the previous section that the reduction to DNF is a ready method for determining if an expression is self-contradictory. A minor problem can now be examined: if the expression is not self-contradictory, is there any ready method for determining whether it is tautologous or contingent? This problem is nonexistent, of course, if trapezoids are used in handling an expression, for as soon as any expression is so symbolized its character as a tautology, self-contradiction, or contingency is evident. The problem evanesces almost as rapidly once the four steps given on page 35 are begun, for in Step 1 the Boolean expansion will correspond to all the lines of the truth table. There are still other effective methods for discerning a tautological DNF. A tautology in CNF is as patent as a self-contradiction is in DNF; hence if the expression is converted into CNF its tautological nature must appear. Yet another device would be to contradict the DNF at hand, then convert that contradiction into DNF; if the original be a tautology, its negation must be self-contradictory and the second DNF will reveal this. But with the possible exception of the trapezoid symbolism, every one of these effective methods is laborious. On the other hand, if a tautologous DNF be simplified, there can be no doubt about its character since it will then be a disjunction of two contradictory literals (*cf.* page 30). From these considerations, this problem can well be restated thus: Is there anything about a tautological DNF which makes it easier to simplify than a contingency? The answer is yes; provided only that a reasonable amount of insight is used, a tautological DNF is necessarily easily simplified.

Consider this DNF (which is already as reduced as 4a and 4b can at this point reduce it):

$$\bar{p}\bar{q}\bar{r} \vee \bar{s}\bar{t} \vee \bar{u}\bar{v} \vee psu \vee psv \vee ptu \vee ptv \vee qsu \vee qsv \vee qtu \vee qtv \vee rsu \vee rsv \vee rtu \vee rtv \vee \bar{w}\bar{x} \vee xyz$$

By using $\bar{u}\bar{v}$ and $rsv$ to generate $rs\bar{u}$, this and $rsu$ can be combined (whether by 4b and 4a or by the generative use of 4c) to result in $rs$ which supplants two of the original disjuncts. (This process of generating new disjuncts is quite straightforward. Any two disjuncts having exactly one letter by which they contradict each other can be so combined; what they generate is the conjunction of both expressions minus that letter.) Using $\bar{u}\bar{v}$ and $rtu$ to generate $rt\bar{v}$, $rt$ can be similarly obtained; $rt$ with $\bar{s}\bar{t}$ yields $r\bar{s}$ which, with the $rs$ previously won, yields $r$, thereby supplanting every disjunct containing it and reducing the first disjunct to $\bar{p}\bar{q}$. The DNF at this point stands thus:

$$\overset{1}{\bar{p}\bar{q}} \vee \overset{2}{\bar{s}\bar{t}} \vee \overset{3}{\bar{u}\bar{v}} \vee \overset{4}{psu} \vee \overset{5}{psv} \vee \overset{6}{ptu} \vee \overset{7}{ptv} \vee \overset{8}{qsu} \vee \overset{9}{qsv} \vee \overset{10}{qtu} \vee \overset{11}{qtv} \vee \overset{12}{\bar{w}\bar{x}} \vee \overset{13}{xyz} \vee \overset{14}{r}$$

The next steps are to combine disjuncts 3 and 8 to generate $qs\bar{v}$, this and 9 to generate $qs$; 3 and 10 produce $qt\bar{v}$, which with 11 yields $qt$, which with 2 produces $q\bar{s}$. This, with the $qs$ previously won, yields $q$ which strikes all disjuncts from 8 onward (save 12, 13, and 14) and reduces the first disjunct to $\bar{p}$. Further simplification is left to the reader as an exercise.

---

To Transform an Expression into DNF, rid it of any

| | | |
|---|---|---|
| 1. horseshoes or equivalences | by | TH and TE, |
| 2. curls | by | TM, |
| 3. dots over wedges | by | TD. |

4. SIMPLIFICATIONS can be effected (even on DNFs within the expression being transformed) by TS.

---

## EXERCISE II–4

1. Transform each of the following into DNF and state whether it is self-contradictory or not.

   a. $p\bar{p} \supset q$ c. $pp \equiv r$ e. $pr \vee p\bar{s} \supset: pq \supset. qr \vee q\bar{r}s$

   b. $q \supset p\bar{p}$ *d. $p \supset q .\equiv r$ f. $pq \supset p \equiv: \bar{p} \supset. \bar{p} \vee r$

2. Using the schemata 4a and 4b reduce the expression (5) at the end of Section 3 to (6) on page 32.

3. Each of the following trapezoid symbols represents a conjunction of literals. Write that conjunction for each.

   a. — x  e. / x x x

   b. x — x x  f. x x x \

   c. x x \ x  g. x x ‾ x x x x x

   d. ‾ x  h. _ x

4. For each of the following expressions, list all its prime implicants. Start with single lines and proceed with successive doublings into related lines until a prime implicant is arrived at.

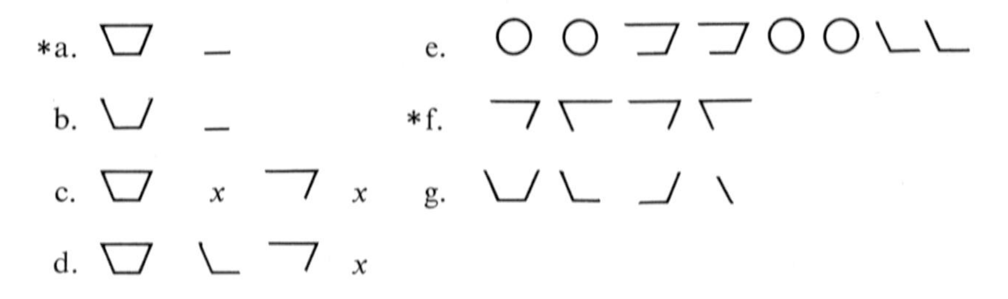

5. With the help of the prime implicants you listed in 4 and the trapezoid symbols themselves, write out a simple DNF for each of the expressions.

*6. Use the effective method for selecting the primes for the simple DNF of the expression in question 4d.

7. Use any of the TS schemata (including the generative use of 4c) to simplify $pq \vee \bar{p}r \vee \bar{q}r$ to $pq \vee r$.

8. Which, if any, of the expressions in question 1 are tautologous?

9. By recourse to the schemata only, simplify these DNFs:

   a. $pq \vee rs \vee \bar{p}\bar{r} \vee \bar{p}\bar{s} \vee \bar{q}\bar{r} \vee \bar{q}\bar{s} \vee t$

b. $pq \vee pr \vee \bar{p}\bar{r} \vee \bar{p}s \vee \bar{q}\bar{r} \vee r\bar{s}$

c. $pq \vee \bar{p}r \vee \bar{q}s \vee \bar{r}\bar{s} \vee \bar{q}t \vee r\bar{t} \vee q\bar{u} \vee \bar{r}u$

10. The text remarks that a tautology in CNF is as patent as a self-contradiction in DNF. Explain why this is so. (Another look at the middle of page 35 may be of help.)

11. In the previous exercise (II–2, page 27) the equivalences adduced under Steps 1, 2, and 3 of *Transformations* were established. In the same way, establish each of the TS schemata — 4a, 4b, and 4c.

12. a. By the use of trapezoids, verify whether (1) on page 30 has remained unchanged by its transformation into (6) on page 32.
    b. What must be the trapezoid symbol for each of the lines 2, 3, 4, and 5 of the transformation on page 31?

## 5. Deduction

It is now plain that the same logical statement can be expressed in many ways. Not only is a DNF the equivalent of the statement giving rise to it, but so is each statement made along the way. To say that each of these statements is an equivalent of the others is to acknowledge that any differences lie only in the writing. In this respect, the truth table — or better yet, the trapezoids — afford the proper basic language. Of the wide variety of possible conventional writings of a statement, preference is given to the DNF because of its greater visibility, so to speak; its standardized form is perhaps a substitute of sorts for the truth table itself. The advantage of simplicity lies all with the truth table or the trapezoids. The indispensable advantage of writability is what the conventional symbolism affords.

But more than this is to be observed in the previous section. It is possible, even while retaining the ordinary symbolism, to demonstrate that if one statement be true another one is, too. For the transformation of a statement into DNF can be considered a step-by-step proof that the original statement is equivalent to the evolved one, the successive rewritings affirming that the original is equivalent to Rewrite #1, which is equivalent to Rewrite #2, etc. This step-by-step way of rewriting is the essence of the deductive technique. It rests on the rudimentary principle that if the premises of an argument can by such successive transformations issue in the argument's conclusion, then that argument is valid.

Each line of such a deduction, it is clear, must be written in accordance with perspicuous rules. For this reason some definitions are in order:

1. By *expression* is meant any well-formed formula. (This is the meaning of this word throughout the previous sections.)
2. An expression is *under* the operator, written or unwritten, most immediately affecting it, as in negation, or of which it is immediately a component

part, as in the other operators. It is also under any operator which in turn affects, or has as a component part, the larger expression constituted by the first-mentioned operator. It is further under any operators affecting these, etc. This is but another way of referring to the hierarchy of operators discussed on page 11.[11]

3. An expression is *asserted* if and only if it is not under a negation, a wedge, a horseshoe, an equivalence, or a quantifier.[12]
4. A *proof*, or deduction, consists of a series of lines, each comprised of one or more assertions and each being either a premise or a regular (*i.e.*, according to rule) rewriting of some earlier assertion(s). There are four rules.

## I. Transformations

*Any assertion may be rewritten with either the whole or any constituent expression replaced by its equivalent.*

This rule permits the exploitation of the transformations to which the previous two sections were devoted, the derivations there studied being a special kind of deduction having recourse to this rule only. Without exception, each deduced line of a proof will be annotated to show by what rule it is justified. When the present rule is the justification, and one or another of these transformations justifies the line, the letters introduced earlier will constitute the annotation. *E.g.*, if from this line in a proof:

5. $q \vee r \cdot \bar{q}s \vee pr \vee q$

the following line were written, it would bear the justification shown:

8. $q\bar{q}s \vee qpr \vee qq \vee r\bar{q}s \vee rpr \vee rq$ 5 TD

If, as a part of the process of distributing, the simplifications referred to on page 31 as rudimentary are effected, the above line can be written more briefly:

8. $qpr \vee q \vee r\bar{q}s \vee rp \vee rq$ 5 TD

[11]By modifying slightly the numbering there suggested, the operators under which any given expression lies are clearly determined. If the same numeral be assigned to associated wedges (or associated dots, or associated equivalences), an expression is then under the operator most immediately affecting it and any operator of equal or greater rank all the way up to the highest ranking (#1), except any equal numbers along the way that lie *beyond* some higher-ranking number.

$$\overset{2}{q r} \overset{1}{\vee} p \overset{1}{\vee} \overset{2}{\sim}(q \overset{3}{\vee} r) \overset{1}{\vee .} s \overset{2}{\supset} p \overset{3}{r} \overset{3}{\sim}(q \overset{4}{\vee} s)$$

In this expression the *r* in the middle is under the wedge to its left, the curl, and each #1 wedge, but not under any other *2*'s or *3*'s than those mentioned. The last parenthesis is under a curl, two conjunctions, one horseshoe, and three wedges.

[12]There are other operators which would have to be banned here had they been incorporated into the symbolism we are using. Quantifiers require no consideration at this point; these operators appear in Ch. IV.

Whichever writing is preferred, the following line can subsequently be derived by TS:

9. $q \vee rs \vee pr$ 8 TS

In a like fashion, TH, TE, or TM can be appealed to.[13] Each appeal to one of these transformation schemata is thus an appeal to Rule I. Rule I can be used without reference to any of these schemata, however; *e.g.*, from these two former lines in a proof:

4. $p \vee q .\equiv r$
5. $s \cdot p \vee q .\supset t$

the following could be justified simply by Rule I itself:

6. $s \cdot r .\supset t$ 5,4 I

The Roman numeral refers to the rule, the Arabic figures to the lines on which the present line is founded. The numeral *5* appears first because line 6 is primarily a repetition thereof, benefited by the replacement of the expression $p \vee q$ by $r$ in accordance with the statement in line 4 that they are equivalent.

## II. Inference

Any recourse to the above transformations results in a new writing which is the equivalent of some previous one. But there are other new statements, not the equivalents of any previous ones, which are nevertheless quite legitimately written and which serve greatly to shorten many proofs. For instance, if $p$ be stated, then it certainly *follows* that *either p or q* can be stated despite the fact that the two statements are not equivalents. And if both $p$ and $q$ are true, then clearly $p$ can be asserted by itself. These two *implicative* inferences are so important that they merit being adopted as schemata:

a. $p$ implies $p \vee q$
b. $pq$ implies $p$

These schemata justify inferences such as these (suppose the first line in each case to be one already existing in the proof):

4. $p\bar{r}s$
5. $p\bar{r}s \vee . \bar{q}t \equiv r$ 4 a

[13]Further schemata might be added in the name of greater rigor, *e.g.*, $pq \equiv qp$ and $p \vee q .\equiv q \vee p$. But as associativity and commutativity have already been exploited in the previous transformations, and as overt manipulation of them tends to encumber a deduction with trivia, they will continue to be exploited as before, *i.e.*, implicitly.

The fact that an equivalence is a biconditional (*cf.* p. 9) allows the inclusion of this pattern:

$$p \equiv q \equiv . p \supset q \cdot q \supset p \qquad \text{TE}$$

This was omitted from the earlier TE schemata (p. 29) because it is not required for reductions to DNF, but it can profitably be kept in mind when writing deductions.

In line 5, line 4 is repeated with a disjunction authorized by schema *a*. Note that this disjoined statement can be any expression whatever. Here is an inference justified by schema *b*:

7. $p\bar{r}s$
8. $\bar{r}$ 7 b

Each of these two inferences is grounded in a single line. It is possible for two lines to yield an implied line:

c. $p$ and $p \supset q$, both being asserted, imply $q$

There is nothing wrong with the assertions corresponding to $p$ and to $p \supset q$ making their appearance in the same line:

3. $\bar{s}t \cdot \bar{s}t \supset r$
4. $r$ 3 c

but this schema also comprehends an inference such as this (take the first two lines to be already asserted):

1. $qs \vee r .\supset p \equiv \bar{q}$
2. $qs \vee r$
3. $p \equiv \bar{q}$ 2,1 c

By adding two variations of schema *c*, an economy in writing lines can be achieved.

c′. $p$ and $\bar{p} \vee q$, both being asserted, imply $q$
c″. $p$ and $\bar{q} \supset \bar{p}$, both being asserted, imply $q$

Each of these is quickly seen to be redundant, for if the two assertions required in *c*′ be given, $q$ can be deduced simply by using TH and *c*. Note the inference on the left:

| | | | |
|---|---|---|---|
| 4. $p$ | | 4. $p$ | |
| 5. $\bar{p} \vee q$ | | 5. $\bar{p} \vee q$ | |
| 6. $p \supset q$ | 5 TH | 6. $q$ | 4,5 c′ |
| 7. $q$ | 4,6 c | | |

But the inference on the right saves one line. Similarly, the schema *c*″ obviates two such appeals to TH:

| | | | |
|---|---|---|---|
| 4. $p$ | | 4. $p$ | |
| 5. $\bar{q} \supset \bar{p}$ | | 5. $\bar{q} \supset \bar{p}$ | |
| 6. $q \vee \bar{p}$ | 5 TH | 6. $q$ | 4,5 c″ |
| 7. $p \supset q$ | 6 TH | | |
| 8. $q$ | 4,7 c | | |

These economies, though small ones, make the inclusion of these schemata worthwhile because they will be resorted to quite frequently.

What is called the Hypothetical Syllogism appears in two of its traditional forms in these schemata: $c$ is *Modus Ponens*, $c''$ is *Modus Tollens*. Schema $c'$ is the Disjunctive Syllogism.

These five implicative schemata form the basis of Rule II, but the Rule itself greatly extends their applicability:

> *In rewriting an expression, any expression therein over which stand only horseshoes to its left, dots, or wedges may be replaced by whatever could be regularly deduced therefrom if it were asserted.*

Without the extension afforded by Rule II, schema $a$ authorizes the following rewriting:

3. $p \supset rs$
4. $p \supset rs\ .\vee\ qt$      3 a

By exploiting Rule II, the following inference is equally licit:

3. $p \supset rs$
4. $p \supset .\ r \vee qt \cdot s$      3 a

In this the $r$ of line 3, qualifying as it does under Rule II, is replaced by $r \vee qt$ (which is deducible from it on the hypothesis that $r$ is asserted). Another deduction which exploits the rule:

6. $p\bar{q}r \vee qs \cdot \bar{u} \supset sw$
7. $r \vee qs \cdot \bar{u} \supset s$      6 b

Notice that the $p\bar{q}r$, one of the expressions on which an inference by schema $b$ was effected, qualifies for the inference; although it has a horseshoe to its right, it is not under that horseshoe. The other expression to which schema $b$ was applied, $sw$, is under a horseshoe but that horseshoe is to its left. It should be noted in this connection that although the rule disallows an inference such as

2. $\sim(pqr)$
3. $\sim(pq)$      2 b (fallacious)

there is no impediment to this inference:

2. $\sim(pqr)$
3. $\sim(pqr) \vee \bar{q}r$      2 a

Here the inference is effected on an expression which, though it *is* a negation, is not *under* one.

This illustrative proof will reward careful examination:

1. $\bar{p} \vee . q \supset \bar{r}$
2. $s \vee pr$ $\quad\therefore q \supset s$
3. $s \vee . q \supset \bar{r} \cdot r$ $\quad$ 2,1 c′ (*i.e.*, a rewriting of line 2 with the invocation of line 1 and schema $c'$. In this line the $p$ of line 2 has been replaced by what it would imply — with the help of line 1 — were it asserted, $q \supset \bar{r}$. The punctuation of line 3 preserves the order of operators in line 2.)
4. $s \vee \bar{q}$ $\quad$ 3 c″ (Here again if the $q \supset \bar{r} \cdot r$ of line 3 were asserted, $\bar{q}$ could be deduced.)
5. $q \supset s$ $\quad$ 4 TH

### III. Reassertion

*Any assertions appearing in previous lines can be conjoined, or simply repeated.*

This rule not only authorizes such inferences as this (suppose lines 1 and 2 to be given):

1. $ps$
2. $r \vee q$
3. $r \vee q \cdot ps$ $\quad$ 2,1 R

but also, by virtue of Rule II, this inference:

4. $prs \vee q$ $\quad$ 2,1 R

for if $r$ of line 2 were asserted, $prs$ could be deduced by Reassertion from line 1 and that $r$. Observe that the code for this rule is *R*. The phrase *or simply repeated* at the end of the rule has no application here but accommodates a rewriting occasionally required in the quantificational logic to be studied in Chapter IV.

### IV. Gratuitous Premise

*A tautology having the form* $p \vee \bar{p}$, $p \supset p$, *or* $p \equiv p$ *may be written as a line.*

The code for this rule will be *GP*. If a complex expression occupies the place of $p$ in the above patterns, its negation or consequent or equivalent may be a transformation. *E.g.*:

$pq \vee \bar{r} \vee . \bar{p} \vee \bar{q} \cdot r$ $\quad$ GP TM

$pq \vee \bar{r} \supset . \sim(pq) \supset \bar{r}$ $\quad$ GP TH

Of course the premise may be adduced without such a transformation:

$pq \vee \bar{r} . \equiv pq \vee \bar{r}$ $\quad$ GP

A final example illustrates various rules:

If Paul parks in front of the driveway, Quincy will evict him. If Ralph loses his bet, Susie will sue. But either Ralph loses his bet or Paul parks in front of the driveway. Hence, if Tom telephones, then either Paul parks in front of the driveway and Quincy evicts him or Ralph loses his bet and Susie sues.

| | | |
|---|---|---|
| 1. | $p \supset q$ | |
| 2. | $r \supset s$ | |
| 3. | $r \vee p$ | $\therefore t \supset . pq \vee rs$ |
| 4. | $pq \vee \bar{p} \vee \bar{q}$ | GP TM |
| 5. | $pq \vee \bar{p} \vee \bar{p}$ | 4,1 c″ |
| 6. | $pq \vee \bar{p}$ | 5 TS |
| 7. | $pq \vee r$ | 6,3 c′ |
| 8. | $pq \vee rr$ | 7 TS |
| 9. | $pq \vee rs$ | 8,2 c |
| 10. | $\bar{t} \vee pq \vee rs$ | 9 a |
| 11. | $t \supset . pq \vee rs$ | 10 TH |

In this proof the gratuitous premise was suggested by the conclusion; note that it provides a statement which by successive rewritings can be converted into the conclusion. Practice will provide insight into which GPs offer promise. Insight and creativity are required inasmuch as the writing of such deductive proofs is obviously not an effective procedure. The rules that pertain to and constitute the Deductive System are not the sort that tell one what to do, only what one *may* do if he chooses.

### Rules of the Deductive System*

I. *Transformation* allows any expression to be replaced by its equivalent; embraces TH, TE, TM, TD, and TS. *I* is also a code if the equivalence is in a prior line.

II. *Inference* allows any expression which is under only
- horseshoes to its left,
- dots, or
- wedges

to be replaced by whatever could be deduced were it asserted.

Its five schemata provide its codes:
- a. $p$ implies $p \vee q$
- b. $pq$ implies $p$
- c. $p$ and $p \supset q$ imply $q$
- c′. $p$ and $\bar{p} \vee q$ imply $q$
- c″. $p$ and $\bar{q} \supset \bar{p}$ imply $q$

III. *Reassertion* provides for conjoining previously appearing assertions. The code is *R*.

IV. *Gratuitous Premises* are not derived from previous lines. They must follow one of three forms: $p \vee \bar{p}$, $p \supset p$, or $p \equiv p$. The code is *GP*.

*A chart summarizing these rules also appears at the back of the book.

## EXERCISE II–5–A

1. Justify each of the deduced lines in the following proofs:

a.
1. $p \vee q \,.\!\supset \bar{r} \supset \bar{s}$
2. $s \vee t \,.\!\supset p\bar{r} \quad \therefore \bar{s}$
3. $\sim(p \vee q) \vee r \vee \bar{s}$
4. $\sim(s \vee t) \vee p\bar{r}$
5. $\bar{s}\bar{t} \vee p\bar{r}$
6. $\bar{p}\bar{q} \vee r \vee \bar{s}$
7. $\bar{p} \vee r \vee \bar{s}$
8. $\bar{s} \vee p\bar{r}$
9. $\bar{s} \vee \sim(\bar{p} \vee r)$
10. $\bar{s} \vee \bar{s}$
11. $\bar{s}$

b. To prove: $p \equiv . \, p \vee pq$
1. $p \vee \bar{p}$
2. $pp \vee \bar{p}\bar{p}\bar{p}$
3. $p \cdot p \vee pq \,.\!\vee \bar{p}\bar{p} \cdot \bar{p} \vee \bar{q}$
4. $p \cdot p \vee pq \,.\!\vee \bar{p}\sim(p \vee pq)$
5. $p \equiv . \, p \vee pq$

c.
1. $r \vee s \,.\!\supset pq$
2. $p \vee \bar{t} \,.\!\supset uv$
3. $x \vee \bar{u} \,.\!\supset ry \quad \therefore u$
4. $\sim(r \vee s) \vee pq$
5. $\bar{r}\bar{s} \vee pq$
6. $\bar{r} \vee p$
7. $\sim(p \vee \bar{t}) \vee uv$
8. $\bar{p}t \vee uv$
9. $\bar{p} \vee u$
10. $\sim(x \vee \bar{u}) \vee ry$
11. $\bar{x}u \vee ry$
12. $u \vee r$
13. $\bar{r} \vee u$
14. $u \vee u$
15. $u$

2. Make the following deductions. After completing each solution you will do well to check each line to make sure it is duly justified. Take special care not to effect an *inference* on an expression that fails to be qualified, Make sure, too, that each line is properly punctuated.

a. 1. $pq \vee \bar{r} \quad \therefore p \vee \bar{r} \vee t$

b. 1. $\bar{p}\bar{q}r \vee s \quad \therefore \bar{t} \,.\!\supset p \supset s$

c. 1. $pq \supset . \, r \vee s \quad \therefore \bar{s} \supset . \, \bar{p} \vee \bar{q} \vee r$

d.
1. $p \vee q$
2. $\bar{q}\bar{r}$
3. $\bar{s} \vee \bar{t} \quad \therefore t \supset p\bar{s}$

e. 1. $p \quad \therefore \bar{p} \supset q$

f.
1. $p \vee q$
2. $\bar{q}\bar{r}$
3. $\bar{s} \vee \bar{t} \quad \therefore \bar{p} \supset \bar{s}p$

g.
1. $p \vee q$
2. $\bar{q}\bar{r}$
3. $\bar{s} \vee \bar{t} \quad \therefore t \supset p\bar{s} \cdot s \supset \bar{t}$

## EXERCISE II–5–B

1. Write a deductive proof of the argument in 2b, page 21.
2. Do the same for 2a of the same exercise.

*3. Write two proofs for the coral snake argument (page 26), one to use a gratuitous premise, the other without it.

4. Prove the argument about Washington, page 21.

*5. Either this report is wrong, or else the truck can be manned instantly and the blaze reached in minimum time if and only if the number of firemen on duty at any one time is increased and a more adequate alarm system is developed. The reliability of the report has been established, and a more adequate alarm system is now in operation. So if the blaze is not reached in minimum time, then the number of firemen on duty at any one time has not been increased.

6. Write proofs for these arguments:

   a. 1. $p \vee qr$
      2. $p \supset r \quad \therefore r$

   *b. 1. $q \vee r .\supset p \supset s$
      2. $\bar{s} \vee t .\supset pq \quad \therefore s$

   c. 1. $p\bar{q} \vee q\bar{r} \vee \bar{p}q \vee \bar{q}r \quad \therefore \bar{p}r \vee p\bar{q} \vee q\bar{r}$

7. The proof on page 49 is by no means as short as it can be made. Line 9 ($pq \vee rs$) can be reached on line 5. Perhaps you can do this without any of the following hints. At any rate, see how few you require.
   1. No GP is needed.
   2. Line 4 is a TS.
   3. A commutation does not require a separate line.
   4. As long as a line is justified by only one rule, more than two lines may be appealed to.
   5. Line 5 so exploits lines 4, 1, and 2.

8. Supposing that an argument has but one premise and the only rule appealed to in the course of a proof is Rule I, would it be possible to write the proof in reverse, *i.e.*, write a proof deriving the premise from the conclusion? Why or why not?

9. Because equivalence is a biconditional or double implication (*cf.* page 45n), a convenient way of proving any equivalence is to deduce each side from the other. By two such deductions prove each of the following:

   a. $p \cdot p \vee q .\equiv p$
   b. $pq \vee \bar{q}r \vee \bar{p}r .\equiv pq \vee r$
   c. $prs \vee pqrs \vee \bar{p}\bar{q}rs \vee qrs .\equiv rs$
   d. $\bar{p}rs \vee pqrs \vee \bar{p}\bar{q}rs \vee \bar{q}rs \vee pq\bar{s} \vee r\bar{s} .\equiv pq \vee r \cdot \bar{s} \vee rs$

*10. If some line of a proof appeals to Rule II, what will be the difference between the trapezoid symbol for the conjunction of the premises and that for the conclusion?

11. When a statement is to be proved and no premise is given (this is often called a *null-premise argument*), how must the first line of the proof be justified?

12. Prove the following tautologous statements by deduction:

*a. $pq \vee r \mathbin{.}\!\supset p \vee r \vee s$

b. $\bar{p}r\bar{s} \vee . \bar{q}r \vee \bar{r} \cdot s :\supset r \vee s$

c. $r \cdot p \vee \bar{r} \vee \bar{s} \supset . \sim(\bar{p}\bar{q}\bar{r}\bar{s}) \supset r \cdot ps \vee \bar{s}$

13. a. According to Rule II, if $p$ implies $q$ then an expression in which $p$ is a disjunct implies the same expression in which $q$ replaces $p$. Make this statement in symbolized form and then use trapezoids to determine whether it is tautologous, contingent, or self-contradictory.

b. Which of the three must this statement be if Rule II is to be regarded as sound?

*c. Similarly test Rule II on the supposition that $p$ is part of a conjunction; *i.e.*, when $p$ implies $q$, does a conjunction containing $p$ imply the same conjunction with $p$ replaced by $q$?

d. And if $p$ is under a wedge which is under a conjunction?

e. And if $p$ is the consequent of an implication?

f. Try $p$ in some position disallowed by Rule II and test the corresponding statement.

## 6. Some Common Devices in Deduction

The student proficient in the use of the deductive technique will have noticed that once a gratuitous premise is introduced, its antecedent (or one of its disjuncts, or one of its sides as an equivalence) quite commonly recurs in line after line. An alternative device, variously called *conditional proof, subsidiary derivation, conditionalization*, or by still other names, is often used in lieu of the gratuitous premise rule. It consists of introducing any expression at all, whether tautologous or not, as an assumption. This requires no justification — a rule provides for it just as our GP provides for the introduction of a tautology — but it does carry with it an obligation. Since any line later than this assumption may well rest in part thereon, no such line can be adduced without an acknowledgment that an extra premise is being assumed. This acknowledgment consists of some such device as the identation of these subsequent lines, a star to the left of each, or a continuous vertical line to their left. But no proof can very well end with this admission (that an extra premise is assumed) still extant. Accordingly, the situation is redeemed and the admission set aside by writing a line which states that *if* the assumed line be true, then such and such a statement is the case. When this line is written, whatever device was used for acknowledging the presence of the extra premise is also set aside.

This explanation will become clearer if the following two proofs are compared. The one to the left is that at the end of the previous section. To the right is the same proof using the device referred to.

| | | | |
|---|---|---|---|
| 1. $p \supset q$ | | 1. $p \supset q$ | |
| 2. $r \supset s$ | | 2. $r \supset s$ | |
| 3. $r \vee p \quad \therefore t \supset . pq \vee rs$ | | 3. $r \vee p \quad \therefore t \supset . pq \vee rs$ | |
| 4. $pq \vee \bar{p} \vee \bar{q}$ | GP TM | 4. $\bar{p} \vee \bar{q}$ | Provisional Assumption |
| 5. $pq \vee \bar{p} \vee \bar{p}$ | 4,1 c″ | 5. $\bar{p} \vee \bar{p}$ | 4,1 c″ |
| 6. $pq \vee \bar{p}$ | 5 TS | 6. $\bar{p}$ | 5 TS |
| 7. $pq \vee r$ | 6,3 c′ | 7. $r$ | 6,3 c′ |
| 8. $pq \vee rr$ | 7 TS | 8. $rr$ | 7 TS |
| 9. $pq \vee rs$ | 8,2 c | 9. $rs$ | 8,2 c |
| 10. $\bar{t} \vee pq \vee rs$ | 9 a | 10. $\bar{p} \vee \bar{q} . \supset rs$ | End of Assumption |
| 11. $t \supset . pq \vee rs$ | 10 TH | 11. $\sim(\bar{p} \vee \bar{q}) \vee rs$ | 10 TH |
| | | 12. $pq \vee rs$ | 11 TM |
| | | 13. $\bar{t} \vee pq \vee rs$ | 12 a |
| | | 14. $t \supset . pq \vee rs$ | 13 TH |

The important thing to be noted in this comparison is that the two devices — that of GP and that of making a provisional assumption later to be acknowledged as an antecedent (line 10) — really come to the same thing. The proof on the left goes forward under the aegis of the tautology that *pq* is the case or else it is not, in which case $\bar{p} \vee \bar{q}$. Lines 4 through 9 on the right go forward on the assumption that it is not, an assumption answered for later (line 10) by stating that line 9 is true *if the assumption is.* On the left, the possibility that *pq* could all along be true is stated each time the tautology is used. Since the two devices serve the same purpose, the student may well ask why this second one is introduced. The answer is that the provisional assumption figures in so many deductive systems it is well for the reader to be acquainted with it despite the fact that it is barred from the present system. (Its use would incur no difficulties within the propositional logic now under consideration. In the quantificational logic of Chapter IV, however, its use would require further restrictions on one of the operations peculiar to that logic, namely, universal generalization.)

Another reason is that it is a useful introduction to a principle which is fundamental to the next method to be discussed, for one use to which the provisional assumption is sometimes put is to assume that the conclusion is false. The purpose of this is to produce a self-contradiction in the event that the argument is really valid. For in a valid argument the premises imply that the conclusion must be true; hence, to affirm the premises and state also that the conclusion is false is to be self-contradictory. Here is a proof of the same argument as before which exploits the assumption, in line 4, that it is invalid, *i.e.*, that the conclusion is not true.

|  |  |  |
|---|---|---|
| 1. | $p \supset q$ | |
| 2. | $r \supset s$ | |
| 3. | $r \vee p \quad \therefore t \supset . pq \vee rs$ | |
| 4. | $\sim(t \supset . pq \vee rs)$ | Provisional Assumption |
| 5. | $t \cdot \sim(pq \vee rs)$ | 4 TH |
| 6. | $\sim(pq \vee rs)$ | 5 R |
| 7. | $\bar{p} \vee \bar{q} \cdot \bar{r} \vee \bar{s}$ | 6 TM |
| 8. | $\bar{p} \vee \bar{p} \cdot \bar{r} \vee \bar{r}$ | 7,1,2 c″ |
| 9. | $\bar{p}\bar{r}$ | 8 TS |
| 10. | $r\bar{r}$ | 9,3 c′ |
| 11. | $\sim(t \supset . pq \vee rs) \supset r\bar{r}$ | End of Assumption |
| 12. | $t \supset . pq \vee rs : \vee r\bar{r}$ | 11 TH |
| 13. | $t \supset . pq \vee rs$ | 12 TS |

Here all the steps beyond the self-contradictory line 10 are really only beating a dead horse because — letting $QRS\ldots$ represent the conjoined premises of an argument and $C$ its conclusion — if

$$QRS\ldots \cdot \sim C . \supset p\bar{p}$$

can be demonstrated, then plainly the antecedent expression

$$QRS\ldots \cdot \sim C$$

must be false, which is to say that by TH

$$QRS\ldots \supset C$$

is true. In short, the moment the conjunction of premises with the denied conclusion can be shown to generate any self-contradiction whatever, validity is established.[14]

## 7. Cross-outs

The gambit just explained underlies the cross-out technique, which consists of listing the transformed premises as a series of assertions containing only dots, wedges, and bars as operators; adding thereto the negated conclusion,

[14] So ancient is this device that early geometers, including Euclid, used it and so did Aristotle. The Latin names given it were *reductio ad absurdum* and *per impossibile*. The transformation schema $\sim(p \supset q) \equiv p\bar{q}$ subtends this principle since it implies $\sim(p\bar{q}) \supset . p \supset q$, another symbolization of the *reductio*.

transformed to meet the same criteria; and proceeding with a series of cancellations designed to evoke any self-contradiction. If and only if the self-contradiction appears is the original argument valid.

Each validity-testing device so far taken up has its merits and weaknesses. The truth table is simple but cumbersome. The trapezoids have the quadruple merit of mitigating that cumbersomeness, mechanizing the merging of expressions over operators (and consequently reducing errors), offering a visual pattern more meaningful to the eye than the columns of *1's* and *0's* offered by the truth table, and extending the number of variables it is practical to cope with. Like truth tables, they afford but one writing for any group of expressions which make the same statement. Nevertheless, trapezoids are also limited in the number of variables they can conveniently accommodate.

Transformations into DNF, used with insight and skill, are a powerful tool. A large number of variables is no deterrent to their use and a self-contradiction, however complicated, must ultimately reveal its true nature in the DNF. The principal drawback in transformations is the awkwardness resulting from having to treat an extended argument as a single statement.[15]

In the deductive technique, this handicap is overcome, for each premise is ordinarily placed on a line by itself and thereby subject to modification while disengaged from the rest. Furthermore, deduction exploits *inferences* whereas transformations are limited to generating equivalent expressions. Finally, deductions afford a path from premises to conclusion which is accessible to the intuition; this is both rewarding and challenging. But these advantages are gained at the expense of effectiveness. A skilled logician might make the transformations more rapidly than a beginner and anticipate shortcuts that escape the beginner's eye; but the technique of reaching a DNF is nevertheless an effective one, since the beginner will obtain the same results without skill or insight provided he is careful not to make an error. Except for those simplifications of DNFs undertaken by insight, all the methods adduced are effective ones. But deductions are not effective. The proof does not make itself; it has to be made. It is a creative process restricted by rules but without the rules to determine the next step.

The cross-out technique combines several of the advantages of the other methods: it is not embarrassed by a large number of variables, any transformations entailed are effected on one assertion at a time, the deductive operations are rudimentary in their simplicity, *and* (until we reach later chapters) *it is completely effective.*

Truth tables (with which we shall continue to work only in the form of trapezoids) will have to be set aside early in subsequent chapters as inadequate to the tasks there to be met with. Indeed, only the deductive technique and the cross-out variation will suffice to meet all the additional demands of the quantificational logic. Finally, when the new demands make every system

[15]This can often be lessened by negating the conclusion to see if this yields a self-contradictory DNF and by other devices. The best such device seems to be the technique now being approached.

non-effective, cross-outs will be seen to require less intuition than any other technique.

The following argument will serve to illustrate the steps of the present method:

$$\begin{array}{ll} 1.\ p \supset q & \\ 2.\ r \equiv s & \\ 3.\ r \vee pt & \quad \therefore\ \bar{u} \vee w \vee qt \vee rs \end{array}$$

1. Negate the statement to be tested; by transformation rid this negation of any curl.

The negation of any implicative statement is the conjunction of the antecedent (the premises) with the negation of the consequent (the conclusion). The above statement can, of course, be denied thus:

$$p \supset q \cdot r \equiv s \cdot r \vee pt \cdot u\bar{w} \cdot \bar{q} \vee \bar{t} \cdot \bar{r} \vee \bar{s}$$

but allowing the line-by-line arrangement to persist conforms better to later steps. There is here no initial curl, although there would be were the expression denied by placing it inside a parenthesis preceded by a curl. Such a procedure might come to mind first for an expression like $p \supset. r \vee pqs :\equiv pq \supset r$. The second part of this first step simply specifies that such a negation should undergo transformation before the second step is begun.

2. Transform each conjunct of this negation into DNF.
3. Place each DNF on a separate line, except for conjoining any wedgeless disjuncts into one line called the line of assertions (LA). The original statement is tautologous if and only if a contradiction of the order $p\bar{p}$ eventually appears in this line.

$$\begin{array}{lll} & & 1.\ \bar{p} \vee q \\ & & 2.\ rs \vee \bar{r}\bar{s} \\ & & 3.\ r \vee pt \\ \text{LA} & \sim\text{C:} & u\bar{w} \\ & & \bar{q} \vee \bar{t} \\ & & \bar{r} \vee \bar{s} \end{array}$$

As is to be seen by the lines which put these two steps into effect, the only wedgeless disjunct (*i.e.*, the only DNF having no wedge — *cf.* page 28) is $u\bar{w}$. Had there been other wedgeless disjuncts, they would have been *placed in the line marked LA, conjoined* to this one.

4. Conjoin further wedgeless disjuncts to the LA in the first of the following ways and, if need be, in the second:
   a. Strike any wedged disjunct containing contradictory literals or containing a literal contradicted by a literal in the LA.

When any DNF is so reduced to but one disjunct (which is *ergo* wedgeless), conjoin that disjunct to the LA.

b. Should disjunctions remain, no contradiction having been arrived at in the LA and no further cross-outs being possible, the existing LA is ramified by conjoining to it, in turn, each of the disjuncts of some (non-redundant) disjunction. Each alternative constitutes a separate branch of the LA; *each* branch must acquire contradictory literals if the original statement is to be judged tautologous.

Notice that Step 4 requires no rewriting of the lines, but only the cross-outs (strikings) executed on the lines produced by Step 3. Inspection of the example reveals no disjunct containing either a *non-u*, which might be struck by virtue of the $u$ in the LA, or a $w$, which the *non-w* of the LA would justify striking. We are thus required to resort to procedure 4b. Let us select line 1 as the disjunction by which the LA is to be ramified, hoping that it proves to be non-redundant. A slash is adopted as a convention for indicating this branching; the LA will now have this appearance,

$$/\bar{p}$$
$$\text{LA} \quad {\sim}\text{C}: u\bar{w}$$
$$/q$$

and the letter $a$ is placed to the left of the first line to indicate that it has been exploited, or transferred to the LA.

$$\text{a} \qquad 1.\ \bar{p} \vee q$$
$$\text{c} \qquad 2.\ rs \vee \underline{\bar{r}}\bar{s}$$
$$\text{b} \qquad 3.\ r \vee \underline{p}t$$
$$/\bar{p}\ r\ r\underline{s}\ \underline{\bar{s}}$$
$$\text{LA} \quad {\sim}\text{C}: u\bar{w}$$
$$/q$$
$$\bar{q} \vee \bar{t}$$
$$\text{d} \qquad \underline{\bar{r}} \vee \bar{s}$$

The *non-p* in the first branch allows the $p$ of the second disjunct in line 3 to be struck (actually *underlined* only, which permits the letters to remain visible); this allows the transference of the remaining disjunct, $r$, to the LA in that branch. When this is done, $b$ is written to the left of that line. This $r$ now permits the *non-r* of the second disjunct of line 2 to be struck, which means that the now wedgeless disjunct $rs$ can be transferred to the LA (we write $c$ to the left of this line). In this manner the first branch eventually acquires two contradictory literals (underscoring calls attention to them in the branch). Attention is now turned to the second branch, for it, too, must yield a self-contradiction if the original is to be judged tautologous. Were this solution being carried out on a blackboard, it might be convenient at this point to erase all of the earlier underscorings to clear the field for the pursuit of the alternative, $q$, assumed by the second branch.[16] The use of a differently colored chalk or pencil would be another alternative, or the underscorings of the first cross-outs might simply be ignored and the second series of cross-outs carried on by using diagonal lines on the struck parts. In any event a new series of key letters (primed this time, and beginning with $b'$) will be convenient for marking the course of the exploitation of the various lines.

a 1. $\bar{p} \vee q$

d′ 2. $rs \vee \underline{\bar{r}}\bar{s}$

c′ 3. $r \vee p\underline{t}$

$/\bar{p}$

LA $\sim$C: $u\bar{w}$

$/q\ \bar{t}\ r\ \underline{r}s\ \underline{\bar{r}}$

b′ $\underline{\bar{q}} \vee \bar{t}$

e′ $\bar{r} \vee \underline{\bar{s}}$

As the student will discover when he applies this technique, most problems can be solved without ever having recourse to Step 4b, *i.e.*, without any ramification of the LA. Even in those situations where Step 4a is not applicable, ramification can frequently be avoided by incorporating into the LA a (wedged) disjunction and then using it to contradict *any other disjunct* (presumably a conjunction such as *rs*) *contradicted by each disjunct of the disjunction thus used* in the LA. For example, the disjunction $p \vee \bar{q}s$ might be incorporated as is into an LA and there serve to strike the first disjunct of a line reading $\bar{p}qu \vee t$,[17] thus allowing the $t$ to be brought into the LA and the solution to be carried on from there. Although this procedure is not included in the above statement of the steps, the student is to use it freely for its practical advantage. It is omitted in the formal explanation because it is, strictly speaking, dispensable, *i.e.*, the procedure 4b (which is *in*dispensable) achieves whatever might be achieved by this procedure. Furthermore, its omission leaves unencumbered the theoretical examination of the cross-out technique which is the subject of the next section.

[16]Had the unbranched portion of the LA ($u\bar{w}$) resulted in any cross-outs, these would be allowed to remain since they would be common to either branch.

[17]Note that $\bar{p}qu$ is contradicted both by $p$, the first disjunct of the expression now in the LA, and by $\bar{q}s$, the second disjunct thereof, thus conforming to the italicized part of the present explanation.

### Steps in the Cross-out Method

1. Negate the statement or argument.
2. Transform each conjunct into a DNF.
3. Place each DNF on a separate line, but conjoin all wedgeless disjuncts to form the LA. If there be no wedgeless disjuncts, make a branched LA.
4. Use the LA to strike contradicting disjuncts on other lines, adding to the LA (or to the branch being exploited) any disjuncts thus made wedgeless.

*The statement is tautologous (or the argument valid) if and only if each branch of the LA shows a self-contradiction.*

## EXERCISE II–7–A

1. Use the cross-out method on question 6, Exercise II–5–B, page 51.
2. Work question 2 of Exercise II–5–A, page 50, using cross-outs.

## EXERCISE II–7–B

1. Use cross-outs on the coral snake problem, page 26.
2. Do the firemen argument, page 51.
3. Work problem 12, page 52.
4. Is the recourse mentioned in the last paragraph of this section applicable to the illustrative problem?
5. Why is there no $a'$ marking any line in the solution of the example?
6. Use cross-outs on this argument:

$$1.\ p \equiv q \ .\!\vee\ \bar{p}q\bar{r}$$
$$2.\ \bar{q}\bar{s} \supset pr$$
$$3.\ \sim(pq\bar{r})$$
$$4.\ \bar{p}s \supset\!.\ q \vee r \qquad \therefore\ p \equiv q \equiv r$$

*7. Write a second cross-out solution to the above problem, this time using a ramified LA if your first solution did not, or vice versa.[18]

8. Sam, who knew the rules governing the Deductive System quite thoroughly, worked for a long time on a solution but failed to deduce the conclusion. "Either this argument is invalid," he said at last, "or else . . . ." Complete his statement.

9. When Sam mastered the cross-out method, he remarked of the same argument, "It's simply invalid." When asked how he could be sure, he gave a brief explanation in which effectiveness figured. Give his explanation.

10. Apply the present method to problems 9 and 10 on page 27.

11. Classify as valid or invalid each of the arguments in question 2, page 21, using cross-outs.

## 8. The Efficacy of Cross-outs

The previous section explained how the cross-out technique is used. The present section will show informally why it is reliable, *i.e.*, that every tautologous statement (by *statement* the special case of *arguments* is subtended) will necessarily be identified as tautologous by the technique and that no non-tautologous statement will ever mistakenly be identified as tautologous. Because the technique deals always with the *negation* of statements under consideration, a more exact expression of what is to be demonstrated is that the technique will generate the self-contradiction of literals mentioned in Step 4 if and only if the negation with which it is concerned really is self-contradictory.

This last allegation is established by our establishing somewhat more yet. It will be shown that the technique merely transforms any negated statement into an equivalent form in which the only manifestation of self-contradictoriness will be the occurrence of contradictory literals in each branch. The reasoning in summary will be this: (1) what results from the application of the technique is a new but equivalent formulation of the same negated statement so that the new formulation is self-contradictory if and only if the original is, and (2) the new formulation is self-contradictory if and only if contradictory literals occur in each branch of the LA.

Consider one by one each separate step of the method. The first step requires no more attention than that given it above. If and only if the statement being tested is tautologous will its negation be self-contradictory. Step 1 provides the negation of the original statement. The remaining steps will reformulate this negation. Note that the transformations called for in Step 1

[18]The disjunction taken into the LA can itself subsequently have some of its disjuncts struck so that it ultimately becomes a wedgeless assertion which may contain one of the contradictory literals of the LA.

and all those involved in obtaining DNFs in Step 2 are such as produce equivalent expressions. This statement can be established by appealing to the truth table (or to trapezoids) in the case of each schematic transformation resorted to (*cf.* Section 3). If each schema is tautologous, then any application thereof will yield expressions equivalent to the expressions so transformed.

The separate lines resulting from Step 3 are to be construed as conjoined; writing them separately is not a means of disguising this fact but of simplifying the method. This is to say that Step 3 is a procedural device, a convention, which allows the writing of:

$$p \vee q$$
$$r \vee s$$
$$pst$$

to signify $p \vee q \cdot r \vee s \cdot p \cdot s \cdot t$. Once this is understood (and the same understanding underlies any deductive technique), Step 3 will be seen to produce a new formulation which is the equivalent of the expression resulting from Step 1 for these reasons: the transformation of a conjunct into DNF results in an equivalent expression; the collection, or conjoining of wedgeless disjuncts, if there are any, into one line effects a conjunction equivalent to that which would obtain were they left on separate lines. The last sentence of Step 3 specifies a mode of procedure. It also constitutes an assertion which will be shown to be true.

Part *a* of Step 4 can be schematized and tested by a truth table. That the right side of the schema is equivalent to the left insures that any operation under 4a results in an equivalent reformulation.

$$\bar{p} \vee q \cdot p .\equiv pq$$

Regarding 4b, an explanation similar to that for Step 3 is in order. The branching of the LA is a procedural convention for representing disjunction. This is clarified by contemplating $\bar{p}\bar{q}$ as a statement to be tested. Its negation (Step 1) is $p \vee q$. When Step 4b is brought to bear on this, it is supplanted by this LA:

```
        /p
LA
        /q
```

However complicated the branching of a line of assertions at the completion of Step 4, it corresponds exactly to an expression that can be written conventionally. For instance,

```
         /s
     /qr
  p      /t
     /qr̄r
```

can be more conventionally written as $p\colon qr \cdot s \vee t .\vee q\bar{r}r$.

Note finally that Step 4 provides for the elimination of every wedged line by transferring it to the LA, for if a line is reduced by cross-outs to a wedgeless disjunct it is transferred under 4a and if not so reduced, it is transferred under 4b. This is to say that barring an earlier completion of the solution by the appearance of a patent self-contradiction, the pursuit of Step 4 will reduce the statement inherited from Step 3 to an equivalent statement in the form of a line of assertions.

From what has preceded, part of what was to be shown follows — that what results from applying the technique is a new but equivalent formulation of the negated statement resulting from Step 1, so that the new formulation, the LA, is self-contradictory if and only if the original is. The remainder of the proof — that this LA is self-contradictory if and only if contradictory literals occur in each branch — rests on the obvious fact that any branch must be comprised solely of literals inasmuch as at no point in the method is anything allowed to constitute, or to be transferred to, the LA unless it be a wedgeless disjunct of a DNF, *i.e.*, a literal or a conjunction of literals. Just as any DNF is self-contradictory if and only if every disjunct thereof is self-contradictory, so any branched LA must be self-contradictory in each branch (*cf.* the second paragraph prior to this). If a self-contradictory LA consists of only one line, it is obviously self-contradictory in that line.

From these considerations it follows that any non-self-contradictory branch of the LA specifies truth value assignments which render the negation resulting from Step 1 consistent and the original statement false.

## SUMMARY

The validity of any argument, whatever its content, is tested by examining the general and abstract form of which it is an example. In logic, form is all-important for this reason.

The trapezoid symbols provide a shorthand writing for the columns of the truth table as well as a mechanical method for 'merging over' the operators. The merging eventuates in a single trapezoid symbol by which the statement under examination is recognized to be tautologous, contingent, or self-contradictory.

Transformations are based on a series of tautologous patterns, so that a transformation effected according to any of these will result in a logically equivalent expression, however much the operators may be changed. The disjunctive normal form is a relatively simple, standardized one which is easily read and manipulated.

Deduction is a process whereby lines are combined or modified to yield other lines — eventually the conclusion itself if the argument is valid. Deduction has recourse not only to transformations but also to inferences, *i.e.*, lines are written which, though not equivalent to previous lines, are implied thereby. Four rules systematize deduction. Auxiliary techniques such as conditional proof are good for the student to be acquainted with but need never be resorted to inasmuch as they can do nothing the deductive system presented here cannot do as well.

Cross-outs restore the effectiveness sacrificed by deductive techniques but contrive to keep the advantages which characterize the latter. Although involving transformations and therefore sometimes tedious, they make relatively short work of even the most complicated problems. In the last section of the chapter it was informally shown that this method identifies a statement as tautologous if and only if it actually is that.

# III

# Class Logic

## 1. Class Variables

The compounded statements discussed in the first two chapters are sometimes called truth-functional statements inasmuch as their truth or falsity is a function of their component variables. This is another way of remarking that $p$ stands for a proposition; the name *propositional* logic comes from this concern with propositions.

It might at first appear that logicians could scarcely entertain an interest in any other kind of symbol. This is mistaken. It is very useful to let a symbol stand for a class of things. Just as $p$, $q$, etc., can stand for *Aunt Charlotte is singing*, *the number of firemen has been increased*, and so on, so $a$, $b$, $c$, and other letters from the early part of the alphabet can be allowed to represent *horses*, *cars manufactured in Michigan since World War II*, *centaurs named Wilbur*, etc. The first symbols must assume one of two values: true or false.[1] But these categories are not appropriate to the symbols for classes. Here the simplest features seem to be belonging or not belonging to another class and having or not having members. Strangely enough, these two features can be integrated — as will be seen in the next section. And in the next chapter it will be seen that even propositional *truth* and class *belonging* can be joined.

## 2. How Modern Logic Arose

Class logic came into being in the first place because men wished to investigate such arguments as

All humans are mortal.
All Greeks are humans.
∴ All Greeks are mortal.

[1]Barring a multi-valued logic in which other values may be introduced.

The propositional logic, which can only represent these three statements by $p$, $q$, and $r$, is powerless to establish any link between them. Some sort of nexus does knit the two premises into the conclusion, however.

When Aristotle attacked the problem, he established for all time these four possible relationships between any two classes:

| | |
|---|---|
| 1. One can affirm that all of a subject class is contained in a predicate class, | All horses are mammals. |
| 2. Or that none of the subject class is so contained, | No Sophomores are Juniors. |
| 3. Or that part of the subject class is contained in the predicate class, | Some college students are coeds. |
| 4. Or that part of the subject class is not so contained. | Some Americans are not married persons. |

It might seem that the statement *All mammals are horses* constitutes a new relationship. In the sense that this is a different affirmation from *All horses are mammals*, this is true. In the sense that a new kind of class relation has been introduced, however, it is false; for it appears upon examination to be simply another instance of including all of a subject class (this time *mammals* instead of *horses*) in a predicate class, and this is the first of Aristotle's relationships, or *forms*, as they are called. Note that the first two of these forms may be said to be *universal*, since they treat of *all* of the subject term (whether including it or excluding it from the predicate term). The adjective *particular* is the proper one for the other two forms, which treat of only a part of the subject class. The first and third propositions are affirmative (and are known as the A-form and the I-form, the first vowels in the Latin word *affirmo*) whereas the second and fourth are negative (E- and O-forms, *nego*).[2] Such abbreviations are unquestionably called for so as to avoid having to identify them as universal affirmative, universal negative, particular affirmative, and particular negative.

Aristotle used these four forms as elementary in the class logic. He is also responsible for the notion of *distribution*, which here has a meaning entirely distinct from that in the previous chapter. A term (or class) is *distributed* by a proposition when every member of that class is referred to. Thus in the universal forms the subject terms are distributed; in the particulars they are left undistributed. As any inclusion of a class in another leaves the latter undistributed, and any exclusion of a class (or even part of a class) from another distributes the latter, the predicates of affirmatives are undistributed and those of negatives are distributed. Aristotle's analysis centered largely on his four *figures* of the class syllogism — these are based on the situation in the premises of the subject and predicate terms of the conclusion — but subsequent followers telescoped his findings into rules for validity and

[2]These abbreviations go back at least as far as the Middle Ages.

systematized his reductions of syllogisms into what he called the perfect syllogism.[3]

By taking it for granted that the syllogism is in proper form, *i.e.*, made up of three propositions in the A-, E-, I-, or O-forms, and that only three terms appear, each in two propositions, these rules can be used to test its validity:

1. Either none, or else two, of the propositions may be negatives. In the latter case, one of the negatives must be the conclusion.
2. The middle term (the one common to the premises) must be distributed at least once.
3. No term distributed by the conclusion may have been left undistributed by the premise in which it occurs.

Aristotle also investigated what is called *immediate* inference, as opposed to the *mediated* inference afforded in the syllogism by the middle term. In immediate inference there is but one premise and the conclusion. *Conversion* is one variety of immediate inference; here the subject and predicate terms are simply exchanged (as in going from *All horses are mammals* to *All mammals are horses*). When is the operation valid? Whenever the only one of the above rules that is applicable, the third, is not violated.

*Obversion* consists of changing the form of the proposition from affirmative to negative (or vice versa) without altering its universality or particularity, and substituting for the predicate term its *complement* (the class of all things outside a class is its complement, *e.g.*, *non-horses* is the complement of *horses* and vice versa, *non-a* of *a*, etc.). Thus the obversion of *Some markets are closed* is *Some markets are not non-closed.* Any of the four propositional forms can be validly obverted.

A third device for making immediate inferences is the Aristotelian *Square of Opposition*, by which inferences are made from one to another form of propositions having the same subject and the same predicate. This will appear in the following chapter.

Only about two hundred years ago, a Swiss mathematician named Euler perceived that these relationships of class inclusion could be easily diagrammed. For the A-form he placed one circle inside another.[4] For the E-form he drew two distinct circles. For the I- and O-forms the circles overlapped, but bore markings to signify which of the areas in the diagram was referred to. Such diagrams can be used to investigate not only immediate inferences, but also syllogisms (a third circle must be used for the third term). This explanation would be expanded here had not Venn revised Euler's circles into a much more elegant apparatus.

[3]This telescoping has the advantage of making tests for validity more mechanical but forfeits some of the richness of Aristotle's investigations. The student may find the *Prior Analytics* more mystifying than illuminating at this stage, but such classics as Jevons' *Elementary Lessons in Logic* (first published in 1870) will give him some of the flavor. *The Fundamentals of Logic* by F. M. Chapman and P. Henle (Charles Scribner's Sons, 1933) gives a clear account of the figures and reduction, including the use of the mnemonic verse which commences "Barbara, Celarent . . . ," a fascinating bit of medieval logical paraphernalia.

[4]Which circle then represents the subject class?

Venn (1834–1923) allowed the two circles always to overlap and then bounded the area with a rectangle, thus enclosing what is called the *universe of discourse*. This provided a map of that universe, so to speak, by dividing it into four segments into one or another of which each thing in the world must fall. Were there horses that are not mammals, they would occupy the area marked 1. In 2 are to be found the horses that are mammals; in 3 the mammals that are not horses (or the non-horses that are mammals, which is the same thing); in 4, such things as are neither horses nor mammals

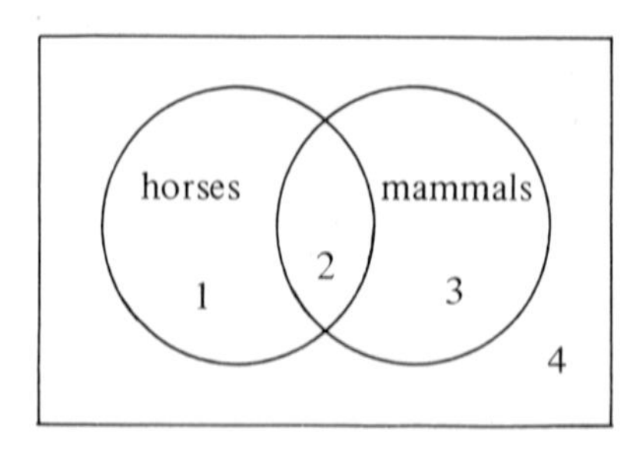


(this text, the square root of 2, etc.). The map offers no statement but affords a matrix on which, by adopting suitable conventions, statements can be made. Venn used shading to indicate that an area has no members, an X to indicate that it has members. To say that *no horses exist* would require the shading of areas 1 and 2 only. To state that *there are horses* requires simply an X, placed on the line between areas 1 and 2 lest we affirm that they are or are not mammals.

This device serves also to relate two classes (thus achieving the integration of the two features of classes mentioned in the second paragraph of this chapter). Suppose *horses* to be the subject class each time and *mammals* the predicate class. An X in area 2 affirms Aristotle's I-form; an X in 1, his O-form; a shaded 2, his E-form; and a shaded 1, his A-form.

Venn's diagram exhibits the validity, or lack thereof, of such immediate inferences as conversions and obversions. These each entail two propositions — the premise and the conclusion. If a diagramming of the premise itself adduces the conclusion, the inference is valid; otherwise it is invalid. Syllogisms are accommodated by introducing a third circle (below the original two and overlapping each) to represent the third class. The principle of testing the argument is the same: the premises are diagrammed; if and only if the conclusion has thereby been diagrammed is the argument valid. In the following example, the shading required by one premise does not allow the X of the other to be placed on the line between areas 4 and 5, but obliges it to be located in area 5; this establishes the conclusion.

Sorites (pronounced *so-rī′-tēz*, singular and plural are alike), which are arguments of four or more propositions (with as many terms), are also testable by Venn diagrams. A sixteen-celled square diagram is more easily

All good pets are well cared for.
Some snakes are good pets.
∴ Some snakes are well cared for.

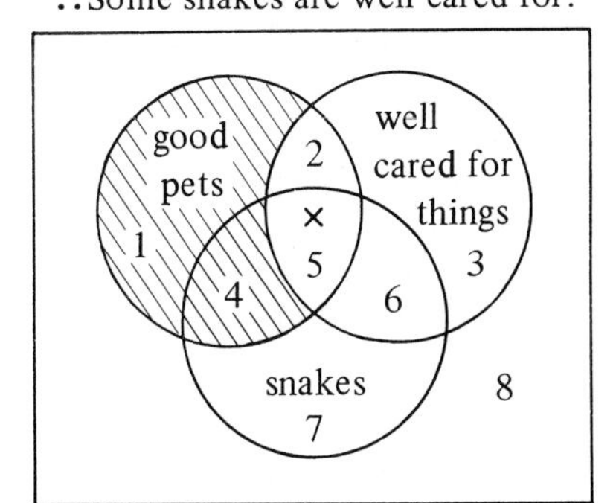


drawn than a fourth overlapping ellipse. A doughnut superimposed over this square successfully provides for a fifth class. Thereafter the best recourse is to draw a second similar figure.

The device of representing classes by letters (and their complements by letters bearing bars) affords a means of converting Venn's diagrams into the algebra invented by George Boole.[5] If $a$ and $b$ are classes, the class of things that are both $a$ and $b$ is variously written $a \cdot b$, $a \cap b$, or simply $ab$ (the last will be used here), and called the *product* of $a$ and $b$, or the intersection of $a$ and $b$. It is read as $ab$. This allows the areas of the Venn diagram to be

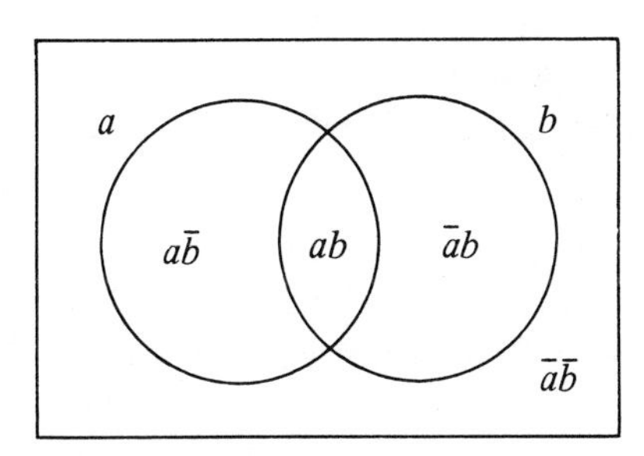


marked in this fashion instead of with numbers. It consequently allows the making of algebraic statements which parallel exactly the diagramming of Venn. As Venn shaded his diagram to indicate a lack of members, Boole equated the class in question to zero, which is used to represent the *null* class, *i.e.*, the class which has no members whatever.[6] That a class has members is properly contrasted with this by making an inequation between the class and the null class.

[5]In Russell's opinion, Boole is the discoverer of pure mathematics. Bertrand Russell, *Mysticism and Logic and Other Essays* (London: George Allen & Unwin, Ltd., 1951), p. 74.

[6]*Centaurs, Fords manufactured before 1895, square circles*, and $a\bar{a}$ exemplify the null class.

| | | | |
|---|---|---|---|
| The I-form: | Some *a* is *b* | is written | $ab \neq 0$ |
| The O-form: | Some *a* is not *b* | is written | $a\bar{b} \neq 0$ |
| The E-form: | No *a* is *b* | is written | $ab = 0$ |
| The A-form: | All *a* is *b* | is written | $a\bar{b} = 0$ |

Valid immediate inferences are represented by Boolean expressions identical to those which represent their premises — identical, that is, except possibly for an inconsequential commutation in the order of the conjoined classes or for the occurrence of a double-barred class in lieu of the class itself.[7]

The adaptation of algebra to the testing of syllogisms is the work of Mrs. Ladd-Franklin and is altogether too simple and neat to be omitted from even this brief account. Once the syllogism is expressed in Boolean algebra, it is converted into an *antilogism* simply by contradicting the conclusion. A Boolean expression is contradicted by changing the equation to an inequation, or vice versa. If the syllogism is valid, this antilogism will be inconsistent, or self-contradictory. Thanks to a theorem of the class calculus,[8] it is provable that a certain pattern of three Boolean expressions constitutes an *inconsistent triad.* Indeed, any inconsistent trio of expressions treating of three classes coincides with that triad as a pattern. Its characteristics are these:

1. There are two equations and one inequation.
2. The term common to the equations is once positive, once negative (*i.e.*, occurs once with a bar).
3. The other two terms are to be found, unchanged, in the inequation.

To determine whether the antilogism be inconsistent (and the original syllogism valid) thereby becomes only a matter of inspection.

One extremely simple modern technique resorts to Aristotle's concept of distribution. By using the numerals 1, 2, and 7 in place of letters to represent the classes in a syllogism and –1, –2, and –7 for their occurrence when complemented, reversing the signs of these terms when they occur undistributed, and adding in 50 for each particular proposition encountered, one will find that a syllogism is valid if and only if the algebraic sum of its premises is equal to that of its conclusion.[9]

[7]$\bar{\bar{a}}$, the complement of the complement of *a*, is the same class as *a*.

[8]*Calculus,* used thus, refers to an axiomatic system. Such systems will be discussed in Chapter V.

[9]Gerald B. Standley, "Two Arithmetical Techniques with Numbered Classes," *Journal of Symbolic Logic*, Vol. 27, #4, 1962, 437f.

### The Testing of Class Arguments

By the rules of *Aristotelian logic* the valid syllogism contains:

1. two negative propositions (one, the conclusion) or else none,
2. a distributed middle term,
3. only such distributed terms in the conclusion as are distributed in the premises.

By *Venn diagrams*, the syllogism is valid if and only if, the premises being entered on the diagram, the conclusion is thereby stated.

*Ladd-Franklin's* test:

1. The antilogism (syllogism with conclusion contradicted) is written in Boolean algebra.
2. The original syllogism is valid if and only if the antilogism shows:
   a. two equations, one inequation,
   b. term common to equations once positive, once negative,
   c. other two terms appearing unchanged in inequation.

*Arithmetical* test: Syllogism uses 1, 2, and 7 for its terms, preceded by a minus sign if the term is complemented. The signs of undistributed terms are changed. To particular propositions 50 is added. Syllogism is valid if and only if the algebraic sum of the premises equals that of the conclusion.

## EXERCISE III–2

1. Label the four parts of a Venn diagram like that on page 68 with class labels, using *h* for horses and *m* for mammals.
2. For each of the following propositions draw a Venn diagram and beside it write the corresponding Boolean expression.
   a. No horses are mammals.
   b. Some horses are mammals.
   c. Some mammals are horses.
   d. All mammals are horses.
   e. Some mammals are not horses.

3. a. Which two of the propositions above are converses of each other? Is this conversion a valid one?
   b. Which two are contradictory? Is there a second contradictory pair? A third?
4. Write the obversion of each of the five propositions in question 2.
5. What relation should the Venn diagrams for the obversions you have just written bear to the diagrams for the original propositions in question 2?
6. Use Venn diagrams to test the validity of each of these syllogisms:
   a. All maples are deciduous.
      No deciduous things are conifers.
      Therefore, no maples are conifers.
   b. No *a* is *b*.
      Some *b* is *c*.
      ∴ Some *c* is not *a*.
7. Test the same two syllogisms by applying Ladd-Franklin's test to their antilogisms.
8. Use the algebraic test (*i.e.*, Ladd-Franklin's) on this syllogism, then test it with a Venn diagram. Explain just what there is about the diagram that establishes the conclusion or leaves it unestablished.

   All senators are loquacious.

   Some loquacious people are mentally deficient.

   Some senators, therefore, are mentally deficient.
9. Use Aristotelian rules for testing the validity of the above three syllogisms. State in each case whether this test agrees with those of modern logic.
10. Without reading the context, attack the syllogism appearing on the following page with all three tools: Venn, Ladd-Franklin, and Aristotelian. If you cannot make the three agree, give up while you are still in a good humor.
11. Apply the arithmetical test to the syllogism in question 10 and then state whether you would classify this technique as modern or Aristotelian and why.

### 3. Existential Import

*Existential import* is a technical phrase. A logic is said to have existential import if it assumes, as does the Aristotelian logic, that the subject classes referred to have members. Modern logic (Venn, Boole, and their successors) makes no such assumption, as is evident from either the Venn diagrams or the Boolean algebra. No reference is being made here to the I- and O-forms; in these existence is expressed in either logic. But in the A-form, *All a is b*, where the classical logic intends that the subject class shall be construed as having members, modern logic makes no such construction but states only that the *a-that-is-non-b* is empty. That there are members in the *a-that-is-b* is not affirmed, and it may well be that $a$ is entirely without members. *All centaurs are red-headed* would not be regarded as true by Aristotle, since there are no centaurs. Modern logic, by contrast, means by the same statement that there are no centaurs that are not red-headed, which is perfectly true.

From these two interpretations, or conventions, flow some curious but wholly consistent results. Aristotle's A-form implies the I-form, which is not the case in modern logic. By the same token certain syllogisms deemed valid by Aristotle are disallowed by modern logic. In view of the difference in existential import, this is not only fitting but indispensable!

All living persons over 200 years of age are blind.
No blind persons are good drivers.
∴ Some living persons over 200 years of age are not good drivers.

Both modern and classical logic agree that the conclusion, expressly stating the existence of living persons over 200, is false. The two logics simply have different ways of accounting for the fact. Classical logic declares the syllogism to be valid but insists that the first premise is nonsense so that there need be no surprise that the conclusion is false.[10] Modern logic has no quarrel with the first premise but points out that the conclusion has no reason to be true since it does not follow from the premises.

## 4. Class vs. Propositional Logic

The operators of the class logic which parallel those of the propositional are striking, to say the least:

| | | |
|---|---|---|
| $ab$ | corresponds to | $pq$ |
| $a \vee b$[11] | corresponds to | $p \vee q$ |
| $a - b$[12] | corresponds to | $p \equiv \bar{q}$ |

[10]Validity of an argument, as was noted on page 19, must not be confused with the truth of its component propositions. There can be invalid arguments made up of true or of false statements, or of any combination of these. There can even be valid arguments made up of false statements or of false premises and a true conclusion. What validity means is that *if* the premises be true, the conclusion must be; hence the only impossible combination is true premises, validity of argument, and a false conclusion.

[11]Also written $a \cup b$ and $a + b$, this symbol stands for the sum of two classes, *i.e.*, everything within either class.

[12]The dash between these variables does not complement $b$ but is an operator. This is the (less used) difference class, the class of things in either $a$ or $b$ excepting those things which are in both. Why this should correspond to the equivalence between $p$ and *non-q* (instead of $q$) is shown in Section 3 of the next chapter.

Even the bar over the class variable may be said to correspond to that over the propositional variable, since it changes the class to what it is not, *i.e.*, an opposite class, in somewhat the same way the truth value of the proposition is changed by the bar.

It is further true that because of these likenesses, certain inquiries can go forward using either notation with equal ease. A case in point is the study of 'logic' underlying relay circuits and the electronic circuits of digital computers.

Despite this, the basic distinction remains. In the above listing of corresponding operators, heed must be given the fact that every expression in the propositional logic is *making a statement*, whereas none of those in the class logic does more than identify a class. When something is *said* of the class $ab$ — that it is empty ($ab = 0$) or that it has members ($ab \neq 0$) — the close similarity evanesces, and these statements are seen to be atomic breakdowns, in terms of $a$ and $b$, of assertions which the propositional logic can only represent grossly as $p$. It was on this note that this chapter opened.

But as surely as the study of logic is not alone a mastery of diverse tools, but also a pursuit of inference *in general*, the task of finding some common language between the two logics continues to invite the inquiring mind. One way that suggests itself is to assign to each area on the Venn diagram a propositional variable. To the area $abc$, for example, let $p$ be assigned, the meaning of $p$ to be that the area has members. *Non-p* will then mean that the same area has no members. With the diagram as a guide, the first premise can be construed as $\bar{p}\bar{r}$, the second as $p \vee q$, and the conclusion as $q \vee u$. Sure enough, the propositional logic bears out the validity of this inference.

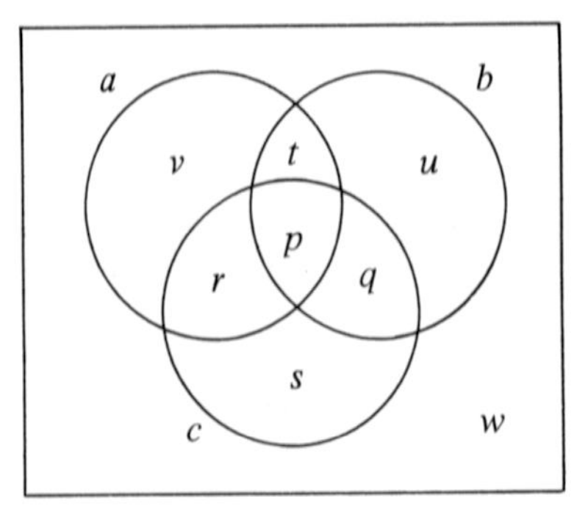


No $a$ is $c$.
Some $b$ is $c$.
∴ Some $b$ is not $a$.

There are three objections to this tour de force. First, the number of classes being $n$, the number of propositional variables that must be available are $2^n$. Second, a considerable number of these variables may actually be called into play. Only four were used in the above transposition, but to accommodate this sorites:

All $a$ is $b$.
No $b$ is $c$.
Some $a$ is $d$.
∴ Some $d$ is not $c$.

no less than eleven of the possible sixteen are actually put to use. But what is much more important is the question of how thorough this transplanting is from class to propositional logic. It is true that the problem is restated and solved in the propositional logic, but without some class logic apparatus such as the diagram, without some way of relating propositional variables to *class products*, the method does not exist. It must be confessed that rather than having transplanted the argument from one to another logic, we have really done nothing more than graft a propositional logic onto the stalk of the class logic. Is there no way, if graft we must, of at least grafting $p$ to $a$, $q$ to $b$, $r$ to $c$, etc.? The answer lies in quantificational logic, a logic as broad as class logic which rests on the propositional and weds them to each other.

## SUMMARY

Classes, as well as propositions, can be usefully symbolized. Indeed, Aristotle's four forms constitute the classical *form*ulation of this logic. His work has been subtended by the modern class logic fathered by Venn, Boole, and others. Both logics are adequate tools not only for immediate inference (two-class inferences) but also for syllogistic (three-class) reasoning in which one term mediates between the other two. The essential distinction between the two logics, apart from the greater breadth of the modern logic, is the matter of existential import.

Certain parallels are to be noted between propositional and class logic. An unyielding difference exists, however, in the true-or-false character of propositions as opposed to the has-members-or-has-not feature of classes.

# IV

# Quantificational Logic

## 1. The Symbols

The I-form of the class logic, *Some a is b*, could almost be put into propositional form by allowing $p$ to signify *Something is in a* and $q$ to mean *Something is in b* and then setting down $pq$. The one difficulty is that there would be no assurance that both propositions were referring to the same thing. For this reason the critical part of what is to be said would be allowed to escape. Any convention overcoming this difficulty would serve. One might draw a circle around the expression, $\textcircled{pq}$, defining the meaning of such a circle to be that every propositional variable within it has the same subject. Or a subscript to each variable might be given the same interpretation. The convention commonly in use is to suffix a letter (from the latter end of the alphabet) by way of assigning the subject. Then as often as that same suffixed index recurs, it is plain that the subject is the same.

What this suggests is

$$px \cdot qx$$

which is correct enough in principle, lacking only some refinements. To understand clearly what refinement is most urgently required, consider the A-form, *All a is b*. Continuing to allow $p$ to stand for *Something is in a* and $q$ for *Something is in b*, the rendition

$$px \supset qx$$

suggests itself, since it reads "If something is in $a$, then that same something is in $b$." The unnerving surprise comes when it is recalled that in the propositional logic $pq$ implies $p \supset q$. It is somewhat less than fitting that the I-form should imply the A-form! There is something about the *something* that still escapes the grasp of this notation.

A return to Aristotle's distinction between universal and particular propositions will right matters. What is not yet incorporated in the notation is that in the I-form there is something — not necessarily everything, but some existing thing — which is both an *a* and a *b*, whereas in the A-form anything at all, whatever it may be, is a *b* provided only that it is an *a*. By attaching appropriate symbols to the propositional statements two ills will be remedied at once: this distinction between something in particular and anything at all will receive its due, and the annoying implication of the A-form by the I-form will be blocked. The conventional means of denoting the universal character of a statement is to prefix the suffix itself to the statement as a whole, enclosing both the one and the other in parentheses.

$$(x)(px \supset qx)^{1}$$

The meaning is that for any $x$ at all, *i.e.*, for every $x$, the statement in the second parenthesis obtains. It is read "For all $x$ (*or* for every $x$), $px$ implies $qx$."

It is immediately plain that the I-form cannot use this universal prefix, for $(x)(px \cdot qx)$ would read, "For every $x$, $px$ and $qx$" which would mean that everything in the wide world is in the class $ab$; whereas the required statement is that there is something in the world that falls in that class. The widely adopted symbol for prefixing a particular statement is '$\exists x$'.

$$(\exists x)(px \cdot qx)$$

This statement is read, "There exists an $x$ (*or* there is an $x$) such that $px$ and $qx$."

From this convention comes the name *quantificational* logic, for the distinction between universal and particular — a distinction of *quantity* in the language of Aristotelian logic — has generated two kinds of *quantifiers:* the universal, $x$, and the existential, $\exists x$.

One more minor imperfection must be eliminated before the symbolism is complete, and this, strangely enough, is the use of the original $p$. For now that it has been assigned a subject, $p$ itself has been reduced to the status of a predicate only. It is well enough to say that $p$ is true and has $x$ for a subject, but this is precisely the same thing as saying that the statement $p$ is true with respect to $x$. Throughout this discussion, for example, $p$ has been standing for *Something* (later made into an $x$) *is an a*. But to say that $x$ is an $a$ is to predicate *a-ness* about $x$, or to assign *a-ness* to $x$, or to make *a-ness* a property of $x$. Indeed, the present logic is known not only as quantificational logic, but also as the *predicate* logic, or as the *functional* logic. In view of this, $F$ and subsequent capitalized letters are commonly used in those places where $p$ and $q$ have been appearing up to now. $F$ stands for *function* just as $p$ stands for *proposition.*[2]

[1]This can be written unambiguously without the first parenthesis: $x(px \supset qx)$. This simpler notation will shortly be adopted. The second parenthesis is worth preserving whenever the scope of the quantifying prefix would otherwise be in doubt.

[2]The $F$ in $Fx$ is called a *propositional function.* To the argument $(x)$ of the function is assigned the function value that is the proposition — precisely the proposition that $p$

$$x(Fx \supset Gx)$$

$$\exists x(FxGx)$$

The symbolism is now full-fledged. It has accrued, incidentally, the capacity to make affirmations of a kind not even considered along the way. If $Fx$ signify, for instance, that $x$ is a farmer, and the statement *Joe is a farmer* is to be made, there is no reason not to set aside the variable $x$ in favor of a *constant*, Joe.

$$Fj$$

For it is not being said that everything is a farmer ($x\ Fx$) nor yet that something is a farmer ($\exists x\ Fx$), but simply that Joe is a farmer: $Fj$. Such renditions are called *singular* statements and use singulars, or constants ($j$ for Joe, in this case) instead of variables. Obviously, a singular is never quantified.

Observe that quantification brings about a new statement. Just as in the propositional logic $p$ and $q$ can be united by an operator so that a new statement emerges as in

$$p \equiv q$$

in which neither $p$ nor $q$ is the assertion but in which their having like truth values — their equivalence — becomes the assertion, so a quantifier makes yet another statement. For instance

$$x(Fx \equiv Gx)$$

makes the statement that for anything in the world, having the property $F$ is equivalent to having the property $G$. All of the propositional operators except negation are *binary*, *i.e.*, connect two statements in some way or other. Negation is *singulary*, operating on only one statement (*cf.* $\bar{p}$ or $\sim(pq \supset r)$). The quantifiers, like negation, modify one statement. Whereas negation reverses the truth value of the statement it subtends, these new symbols quantify, whether universally or existentially, the statement within their scope.[3]

This point is emphasized so that the quantifiers may be clearly recognized as operators, too, and hence seen to participate in the hierarchy of rankings. Of these two statements:

$$x\ Fx \supset x\ Gx \tag{1}$$

$$x(Fx \supset Gx) \tag{2}$$

(1) is an implication; (2) a quantification. The quantifications in (1) are under the implication. The implication in (2) is under the quantification. The

---

represented when $px$ was being used. Alternatively, $F$ can be said to be predicated of $x$, or to be a property of $x$, or (in the terms of Chapter III) to be a class to which $x$ belongs.

[3]The parenthesis enclosing the statement under the scope of a quantifier can, of course, be supplanted by dots: $x.Fx\ Gx$; but the use of parentheses would seem to be preferred here, as in negation, since both quantification and negation are singulary operators.

critical difference in ranking results in an enormous difference in meaning, (1) being read "If everything in the world be an $F$, then everything in the world is a $G$," whereas (2) is "Take what you will in the world, if it be an $F$ then it is a $G$." If anything in the world is not an $F$, (1) is true, its antecedent being false. The occurrence of that *non-F* has no effect on the truth of (2). The student will recall that failure to punctuate properly in the propositional logic was disastrous. He is encouraged to bear in mind that quantificational logic is no different in this respect.

## 2. The Square of Opposition

A very old device in logic is the Aristotelian Square of Opposition:

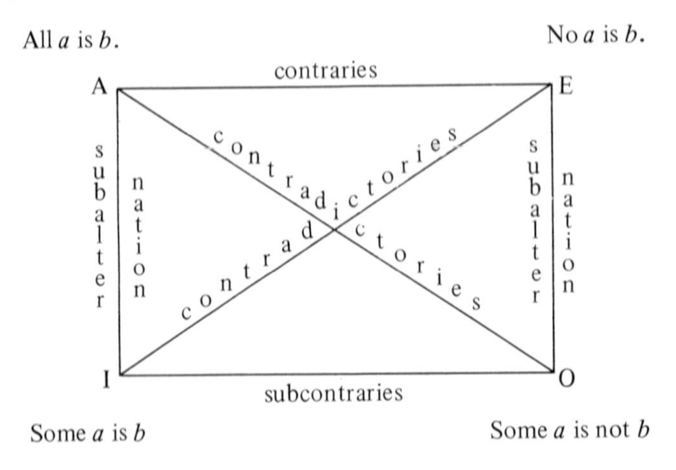


This diagram summarizes the relationships of the four forms to each other. Consider first the contradictories. If it be true that *All a is b*, it can only be false that *Some a is not b* and vice versa. The same mutual exclusivity characterizes the E- and I-form pair. To say that propositions are contradictory, then, is to say that their truth values are necessarily opposite.

The contrariety of the A-form to the E-form is not quite the same thing, since conceivably both forms could be false at the same time (let $a$ be Americans and $b$ be college students, as an example). They cannot, however, both be true at the same time. Subcontrariety is a relationship complementary to this: the I- and O-forms could very well both be true; they cannot both be false. Subalternation is the relationship borne by the subalterns (the lower forms) to their respective universals (above them). The truth of a universal implies that of its subaltern; the falsity of a subaltern, the falsity of its universal.

All of these relationships, except contradiction, rely on the existential import of the Aristotelian logic. If this import is set aside, as it is by modern logic whether in class logic or in quantificational, every one of the relationships on the sides of the square disappears:

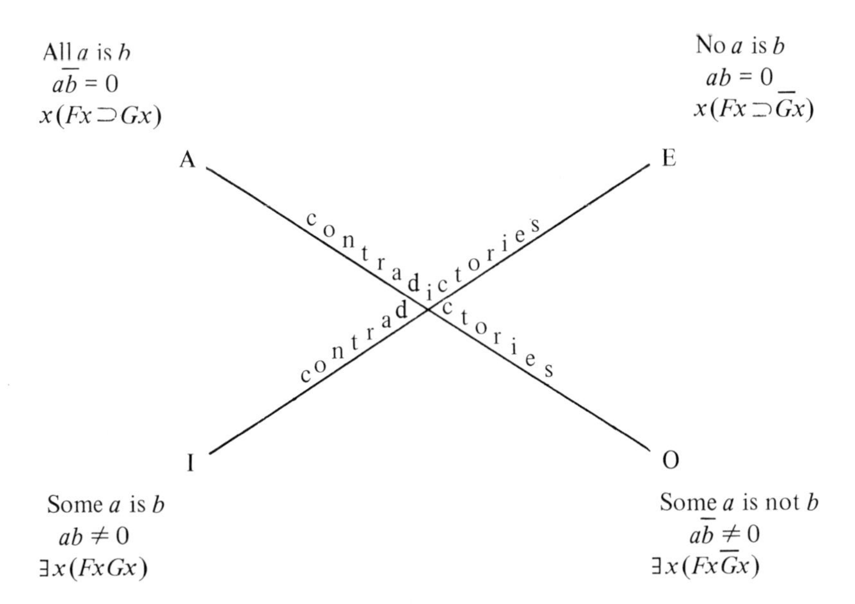


On this modern square there is nothing to indicate that the A- and the E-forms could not both be true. Adducing both statements on a Venn diagram produces no contradiction but rather the affirmation that there are no $a$'s, which occasions no inconsistency since no existential import inheres in either statement. In the Boolean algebra, to affirm both $a\bar{b} = 0$ and $ab = 0$ again amounts to declaring that $a$ is empty, $a = 0$. And in quantificational logic, to say that anything in the world that is an $F$ is a $G$ and that anything in the world that is an $F$ is a non-$G$ is only to affirm that there must be nothing in the world that is an $F$. If $p$ be false, both $p \supset q$ and $p \supset \bar{q}$ are equally true, and conversely, to assert both $p \supset q$ and $p \supset \bar{q}$ is to assert the falsity of $p$.

That the contradictories are still contradictories, however, is plain from a comparison of the statements in question whether on the Venn diagram (on which opposite, incompatible statements must be set down with respect to the same areas), or in the Boolean expressions (where it will be recalled that contradiction consists simply in the substitution of an inequation for an equation or vice versa), or in the quantificational expressions (where the A-form, $x(Fx \supset Gx)$, states that anything in the world that is an $F$ is a $G$, whereas the O-form expressly declares that there exists something in the world that is an $F$ but not a $G$). If either statement be true, the contradictory is necessarily false; if either be false, the other is necessarily true.

From this can be discerned the basis for a transformation indispensable to the manipulation of quantificational logic. Note that the propositional statement in the O-form, $Fx\bar{G}x$, the equivalent of $\sim(Fx \supset Gx)$, is contradicted by $Fx \supset Gx$. Note, too, the meanings they have. It is clear that any time a function is attributed to everything, whatever that function may be, the resulting statement will be contradicted by another affirming the existence of something not having that function; *i.e.*, any statement of the form $x\,Fx$ will be contradicted by the statement $\exists x\,\overline{F}x$. If these be contradictories, there is little difficulty in writing these equivalences:

$$x\,Fx \equiv \sim(\exists x\,\overline{F}x) \qquad \text{TQ}$$
$$x\,\overline{F}x \equiv \sim(\exists x\,Fx)$$

which in turn will lead to these:

$$\sim(x\,Fx) \equiv \exists x\,\overline{F}x \qquad \text{TQ}$$
$$\sim(x\,\overline{F}x) \equiv \exists x\,Fx$$

**EXERCISE IV–2**

1. Place the A-form on a Venn diagram. Has the I-form been affirmed by what you have set down? Does the A-form imply the I-form on Venn diagrams?

2. Does the E-form imply the O-form on Venn diagrams? Explain what the lack of existential import in modern logic has to do with the failure of the universals to imply their subalterns.

3. If everything in the world were *G*'s and some things were *F*'s, would the I-form be true or false? The O-form?

*4. In what kind of universe would both the I-form and the O-form be false?

*5. Transform $\sim(x(Fx \supset \overline{G}x))$[4] into a statement having no curl.

6. Do the same for each of these:

   a. $\sim\exists x\,(\overline{F}x \vee \overline{G}x)$ c. $\sim x\,(Fx \equiv \overline{G}x)$
   b. $\sim\exists x\,(Fx \vee Gx\overline{H}x)$ d. $\sim x\,(Fx \vee \overline{F}x)$

*7. If $x\,Fx$ were the conclusion of an argument, and you wished to negate this conclusion, what could you write that would have no curl?

8. Answer the same question with respect to:

   a. $\exists x\,(FxGx)$ c. $\exists x\,Fx \supset x\,Gx$
   b. $x\,(Fx \supset Gx)$ d. $x\,Fx \vee \exists x\,(GxHx) \vee x(Jx \supset Kx)$

[4]Hereafter (starting with the next question) the parenthesis enclosing a negated quantified statement will be omitted. Thus $\sim(x(Fx \supset \overline{G}x))$ will be written $\sim x(Fx \supset \overline{G}x)$. This is possible because the negation preceding a quantifier can only be construed as negating the statement constituted by the quantifier.

## 3. Trapezoids in the Quantificational Logic[5]

The predicate logic consists of the propositional logic modified in two ways: by the addition of an index which makes the subject of one proposition the subject of another, and by the prefixing of a quantifier for that index. It would seem to follow that some modification of the use of the truth table could be expected to provide a problem-solving technique. This surmise is correct.

In showing this to be true, trapezoids will be used instead of truth tables. The procedure will be described first and explained later. As first presented, it will be applicable to arguments in which each premise (and the conclusion) stands under a single quantifier. The steps are these:

1. Negate the conclusion. The premises, plus this denied conclusion, can now be expected to contain a self-contradiction if the argument be valid.
2. Free each statement of any curl preceding its quantifier.
3. For each of the universal statements represent the statement proper — *i.e.*, the part falling within the scope of the quantifier — in trapezoids and then merge these symbols by conjunction to form a trapezoidal expression which will be called the KU (conjunction of universals).
4. Put each of the statements proper of the existential statements into trapezoids, but keep them separate; each one is to be examined by itself.
5. This step consists of inspecting. The original argument is valid if and only if there appears a self-contradiction somewhere, whether it be
   a. one of the statements themselves that is self-contradictory (empty of lines in the trapezoid symbolism), or
   b. a KU empty of lines, or
   c. some existential statement having no lines in common with the KU.

By way of illustration:

> No prince is a quiet sibling. Any quiet sibling is either resting or else not a prince. There are quiet things that rest. There are quiet things that are either princes that are siblings or else resting though not princes. Hence, some non-prince is resting and quiet.

The argument is symbolized by letting $Px$ mean $x$ is a prince; $Qx$, $x$ is quiet; $Rx$, $x$ is resting; and $Sx$, $x$ is a sibling.

1. $x(Px \supset \sim(QxSx))$
2. $x(QxSx \supset . \ Rx \vee \overline{P}x)$
3. $\exists x(QxRx)$
4. $\exists x(Qx \cdot PxSx \vee Rx\overline{P}x) \quad \therefore \exists x(\overline{P}xRxQx)$

[5]The method explained in this section first appeared in "Testing Singly Quantified Tautologies," by the author, *Journal of Symbolic Logic*, Vol. 31, #3, 1966, 478–480.

The first step adds $\sim\exists x(\bar{P}xRxQx)$ to the list of premises. The second step makes this, with the aid of TQ,

$$\sim C\colon x(Px \vee \bar{R}x \vee \bar{Q}x)$$

The KU of the third step is derived thus:[6]

1. ▽ ▽ ○ ○
2. ○ ▽ ○ ○

~C: ⊏ ○ ⊏ ○

KU: ⌜ ▽ ⊏ ○

The fourth step symbolizes the existentials:

3. ⌟/ x ⌟/ x

4. ⌟/ _ / x

Step 5 (inspection) discovers no self-contradiction under either heading *a* or *b*. Line 3 has one line in common with the KU (the line $pqr\bar{s}$) so that no contradiction exists there, thus indicating, incidentally, that this premise is redundant, *i.e.*, contributes nothing to the argument. Line 4, however, lies entirely outside the KU, which constitutes a contradiction. The original argument is valid.

The most easily intuited explanation of why this technique is sound reverts to the class logic. The connection which the following explanation establishes, by the way, should remove any lingering doubts as to the essential identity between the class logic and this level of quantificational logic.

Consider once more the separate lines of the trapezoid, this time in connection with a Venn diagram:

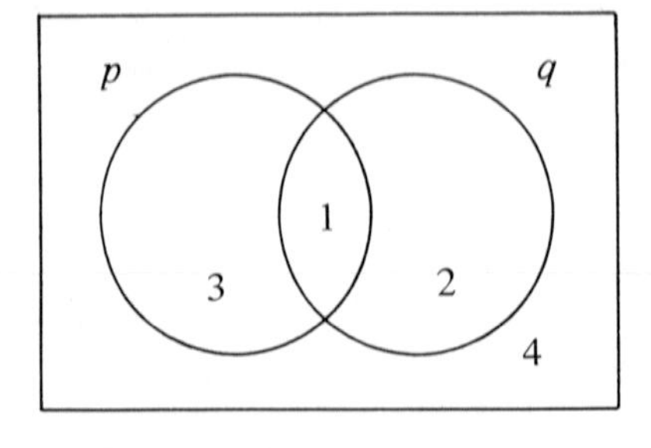


4
3 ▽ 2
1

Here is the device hinted at in the final lines of the previous chapter — a means of relating the propositional to the class logic without an undue proliferation of variables. The difference is that now a distinction has been clearly drawn between universal statements and existential ones.

[6]See footnote 4 on p. 25.

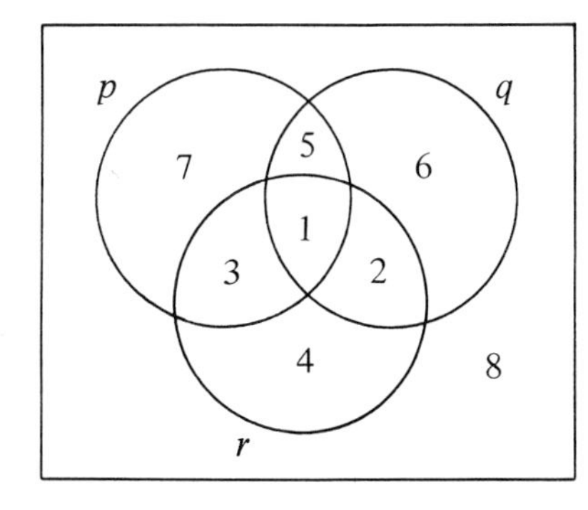


4
3 ▽ 2
1

8
7 ▽ 6
5

What is the function in this technique of the KU? Observe that the universal statement is *universal* in both this and the class logic. In quantificational logic, it bears an $x$ as a prefix. In the Venn diagrams the universal results in emptying (by the convention of shading) specified areas. The trapezoid symbolism for a universal statement thus corresponds line by line to the areas of the Venn diagram which the same statement in the class logic would *leave unemptied.* In other words, the symbol designates those areas on the diagram in which it still remains possible for something to exist. The KU, conjoining as it does the universal statements, shrinks still further the area of the universe of discourse that can have members.

Before going on, consider alternative *b* under Step 5. A KU empty of lines would correspond to a Venn diagram every area of which has been shaded. This would assert that the universe is empty, and therein lies the self-contradiction, since logic presupposes a non-empty universe.[7]

The trapezoid symbol for the existential statement (*particular* statement in the terminology of the class logic) corresponds line by line to those areas *in one or another of which* an X would be placed on the Venn diagrams. It affirms (whether on the diagram or stated in the symbolism of the trapezoid) that one or another of these areas has members. Consider now alternative *c* under Step 5. An existential symbol having one or more lines in common with the KU constitutes nothing impossible, for both statements could be true, *i.e.*, the existence of something in an area represented by such a common line would be in keeping with what the KU allows to exist and yet keep the existential statement from being false. But an existential symbol whose lines lie entirely outside the KU affirms that something exists precisely where the KU declares that nothing exists. This is contradictory.

Singular statements require special handling in order to be made amenable to the present technique. They are artificially converted into existential statements. *Joe is a farmer*, for example, would be treated as *There exists a Joe* (making Joe-ness into a property) *that is a farmer:* $\exists x(JxFx)$. If a second premise declared that Joe is a senator, this would have to be added *under the same quantifier:* $\exists x(JxFxSx)$. Although logically sound, this is altogether

[7]This requires qualification. Some logicians give serious attention to a logic designed to include the case of the empty universe. There is nothing irregular, however, about the present limitation.

artificial and is to be resorted to only in connection with the technique under consideration here. In every other technique, the standard way of expressing singular statements (*e.g.*, $Fj$, see page 79) is to be preferred.

It was stated earlier that this technique is applicable to arguments in which each premise stands under a single quantifier. Its range is really somewhat wider than this, *i.e.*, it can be used with premises in which occurs a second quantifier separated from the first by some propositional operator, *e.g.*,

$$x(Fx \supset Gx) \supset x(Fx \supset Hx)$$

In such a statement, any propositional operator standing over the quantifications must first be transformed into a wedge or a dot. Should the transformation result in a dot,[8] the results can be construed as two statements.[9] Should it result in a wedge, as in the case of the line given

$$\sim x(Fx \supset Gx) \vee x(Fx \supset Hx) \qquad \text{TH}$$

which transforms again to

$$\exists x(Fx\overline{G}x) \vee x(Fx \supset Hx) \qquad \text{TQ}$$

the entire argument becomes in effect two arguments with the alternate premises. And unless a contradiction appears *no matter which premise is used*, the argument cannot be considered valid.

When both quantifications are the same, there is a temptation to merge the trapezoid symbols over the wedge. This is harmless if the quantifiers are existential; otherwise it is an egregious error, for it disregards the importance of punctuation and supposes there is no difference between

$$x(Fx \vee \overline{F}x) \quad (1) \qquad \text{and} \qquad x\,Fx \vee x\,\overline{F}x \quad (2).$$

In (1) the statement is under one quantifier and the symbols for $Fx$ and for $\overline{F}x$ should be merged over the wedge. In (2) two alternative premises are adduced, as stated above. Here is an example of a problem engaging this wider application of the method:

1. $x(Fx \vee Gx \,.\!\supset Hx)$
2. $\exists x(\overline{G}xHx) \supset \exists x(HxFx) \quad \therefore x(Gx \vee \overline{F}x)$

The first two steps add to the list of statements:

$$\sim\text{C}: \exists x(\overline{G}xFx)$$

The second premise, being an implication, invokes the procedure in question. Transformation results in a wedge:

$$2.\ x(Gx \vee \overline{H}x) \vee \exists x(HxFx)$$

[8]The transformation of $\sim(x\,Fx \supset x\,Gx)$ would be a conjunction.

[9]All of the premises of an argument are really conjoined even though they are entered on separate lines. The negated conclusion is also tacitly conjoined in the same way.

so that the list of statements is now divided into two alternative lists, each of which must show a self-contradiction if the argument is to be judged valid. Steps 3 and 4 are carried out on two sets of statements:

| | |
|---|---|
| 1. $x(Fx \vee Gx \,.\!\supset Hx)$ | 1. $x(Fx \vee Gx \,.\!\supset Hx)$ |
| 2a. $x(Gx \vee \overline{H}x)$ | 2b. $\exists x(HxFx)$ |
| ~C: $\exists x(\overline{G}xFx)$ | ~C: $\exists x(\overline{G}xFx)$ |

When ready for inspection, the two look like this:[10]

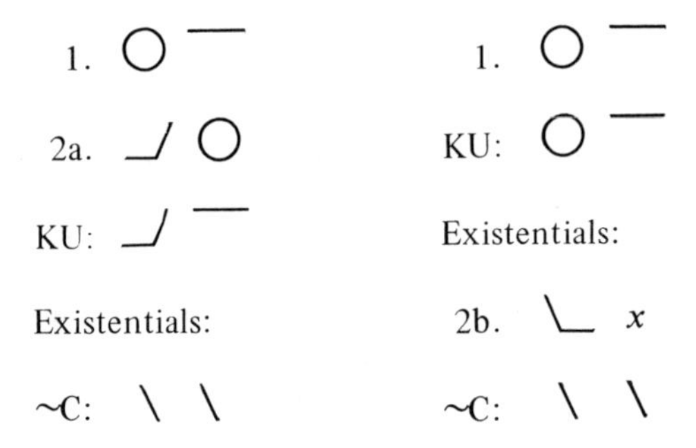

Inspection reveals a self-contradiction in the alternative on the left (between the KU and the only existential), but because no contradiction appears in the second alternative, the argument is finally invalid.

## EXERCISE IV–3

1. Make a Venn diagram for each of the four Aristotelian forms. Beside each diagram, write the corresponding quantified expression. Label the circles on the diagram $F$ and $G$ to simplify matters. Beneath each quantification write the corresponding trapezoid symbol. (Note in each instance the connection between trapezoid and diagram mentioned on page 84.)
2. a. Test this syllogism first by Venn:

   All *a* is *b*.
   Some *c* is *b*.
   ∴ Some *a* is *c*.

   b. Then translate the same argument into the quantificational logic and test its validity with trapezoids.

[10]The function $F$ has been treated as $p$; $G$, as $q$; and $H$, as $r$.

3. On what principle are the numbers assigned to the areas in the diagrams on pages 84 and 85?
4. Determine the validity of these two arguments.

   a. 1. $x(Fx \vee \bar{F}x)$
      2. $\exists x\, Fx \quad \therefore x\, Fx$

   b. 1. $x\, Fx \vee x\, \bar{F}x$
      2. $\exists x\, Fx \quad \therefore x\, Fx$
5. Put into words the first premise of each of the two above problems.
6. The account of why the trapezoid technique works made no reference to possibility *a* under Step 5. Write briefly disposing of this possibility.
7. Is $x(Fx \vee \bar{F}x)$ a tautologous statement? Can you prove your answer with trapezoids? Hint: Treat the statement as a conclusion having no premises. After answering this question, reexamine your answer to question 6 to see if it still satisfies you.
8. a. Is the technique presented here effective?
   b. If no contradiction appears, can one be certain that the argument is invalid?
9. It is implied by the remark in line 20 of page 86 that this statement is logically true:

$$\exists x\, Fx \vee \exists x\, Gx \,.\!\equiv \exists x(Fx \vee Gx)$$

   Test this equivalence, handling it as two implications.

## 4. Working with Bound Variables[11]

The trapezoid method, besides becoming complicated when dealing with premises not under one quantifier (witness the final example), is limited to what are called singly quantified expressions and is therefore inadequate to a large area of the quantificational logic which has not yet been approached. An expression in which one quantifier stands under another is said to be multiply quantified. One example of such an expression is

$$x(\exists y(FxGy))$$

This can be rewritten in singly quantified conjuncts having the same meaning:

$$x\, Fx \cdot \exists y\, Gy$$

or, for that matter,

$$x\, Fx \cdot \exists x\, Gx$$

Other such functions, however, connect two things to each other instead of characterizing only one thing. Just as $Fx$ can be assigned the meaning *x is a father*, $Fxy$ can be used to symbolize *x is father to y*. Such relational predi-

[11]The meaning of *bound variable* will appear in the next section.

cates demand second quantifiers, and functions predicating a connection between three things (triadic functions) or four things (tetradic) commonly require as many quantifiers. To say that every man has a father, for instance, we must write

$$x(Mx \supset \exists y\ Fyx)$$

and that there are librarians who help students find books,

$$\exists x(\exists y(\exists z(LxSyBzHxyz)))$$[12]

These relational predicates are taken up later; they are mentioned here only by way of showing how short the trapezoid technique must fall of accommodating all the demands of the predicate logic.[13]

Hence, more powerful techniques must be resorted to. The method most in use is that of deduction, line by line; the given premises are rewritten in new forms until the conclusion emerges. One way of doing this provides for stripping off the quantifiers so that the remaining expressions, containing as they must only functions of variables (corresponding to the $p$, $q$, $r$, etc., of the propositional logic) and these connected by the operators of the propositional logic, can be handled entirely by techniques of that logic. Provisions are also made for restoring quantifiers according to rule. Considering the cumbersomeness and complexities of quantification, this method can be called elegant if for no other reason than that it works! It further merits the reader's attention because it is in common use. The next sections are devoted to an exposition of such a deductive technique.

The present section is nothing more than a preparatory study. Here the quantifiers are left on the expressions. This procedure is not rigorously organized; it is not strictly a method, as was the trapezoid technique, but rests on insight instead of rules. Its relative simplicity will help the student understand the deductive method to come.

Like the trapezoid technique, it negates the conclusion and probes the validity of the original argument by searching for an inconsistency in the anti-argument.

> No one who has a sore throat can very well be garrulous. Yet when one doesn't have a sore throat he has to listen. Some of my aunts have to listen. Hence, if all of my aunts are garrulous, they all have to listen.

Let $Sx$ mean *x has a sore throat; Gx, x is garrulous; Lx, x has to listen; Ax, x is an aunt of mine.* Below the premises appears the denial of the conclusion.

[12]This surfeit of parentheses is avoidable. As no ambiguity results from writing $\exists x\ \exists y\ \exists z(LxSyBzHxyz)$ for this same expression, the simpler writing will be adopted hereafter.

[13]Even within monadic functions, not every instance of multiple quantification can be readily reduced to a singly quantified expression, *i.e.*, one in which no quantifier is within the scope of another. $x\ \exists y(Fx \equiv Gy)$ can be written in singly quantified form only by considerable transformation.

The original conclusion is disregarded, of course; the self-contradiction must appear in the five lines on the left if the argument is to be judged valid.

*Solution*

1. $x(Sx \supset \overline{G}x)$
2. $x(\overline{S}x \supset Lx)$
3. $\exists x(AxLx)$ $\therefore x(Ax \supset Gx) \supset x(Ax \supset Lx)$ $G\overline{S}L$

$\sim$C: $x(Ax \supset Gx)$

$\exists x(Ax\overline{L}x)$ $S\overline{G}\overline{A}$

Recalling that each universal statement applies to everything in the universe, each existential statement is tested to see if any self-contradiction appears when the universals are all satisfied with respect to it. Consider first line 3. This premise affirms that there exists in the universe an element that is both an $A$ and an $L$. What further characteristics do the universal statements (including the one in the negated conclusion) impose on such an element? The first perusal of lines 1 and 2 shows no such further demands. The next to last line, however, requires that anything that is an $A$ be a $G$; this function ($G$) is therefore added to the characteristics of the $AL$ in line 3 ($G$ may conveniently be written to the right). It now appears from line 1 that $\overline{S}$ must be added, too, which in turn requires by line 2 that $L$ be added. With all this, however, no contradiction is to be seen, *i.e.*, the $AL$ of 3 is also $G$ and $\overline{S}$ (and again $L$ by the time each universal is consulted), but this is altogether consistent.

Not so with the $A\overline{L}$ declared by the last line to exist. This turns out to be further characterized by $S$ (by line 2), $\overline{G}$ (line 1), and $\overline{A}$ (the next to last line). Clearly it cannot be both $A$ and *non-A*, and this contradiction reveals that the original argument is valid. If no inconsistency had appeared from this final examination, this argument would have been classifiable as invalid.

The exercise to follow will reveal that this is anything but a perfected method. It can be formalized (the Appendix does exactly this), but a more customary way of making deductions provides for stripping away quantifiers. The following sections regularize deductions of this sort by subjecting the inference here placed before the intuition to rules designed to disallow any inferences that are not valid.

## EXERCISE IV–4

1. In this procedure an existential is never 'tested' against another existential. Why not?

2. One of the following three arguments is invalid. Using this section's procedures, validate the other two.

   a. 1. $x(Fx \supset Gx)$
      2. $\exists x(FxHx)$ $\quad \therefore \exists x(HxGx)$

   b. Some accountants criticize their profession. But then any discouraged or reflective person will do so. All accountants are persons. But as there are discouraged persons who are not accountants, every accountant must criticize his profession.

   c. There are persons who are millionaires and unhappy. Every millionaire who is philanthropic is happy. Therefore there are unphilanthropic people.

3. Prove each of the following with this procedure, staying as best you can within what you would call acceptable lines. If you need to introduce a principle not discussed in this section, express that principle in words as clearly as you can. Use the standard way of expressing singular statements (*cf.* page 79).

   a. Bill is a person who farms. Any farmer likes to eat. Hence, there is someone who likes to eat.

   b. 1. $x(Ax \supset Bx)$
      2. $x(Bx \supset \overline{A}x)$ $\quad \therefore \exists x\, \overline{A}x$

   c. If Hank is a midshipman, there are midshipmen who are tactless persons. No tactless person can be trusted to escort the Blueberry Queen on her tour. But all midshipmen can be so trusted. So Hank must not be a midshipman.

4. a. Now that you have worked question 3, summarize what you think will have to be provided for when this procedure is replaced by a deductive technique if that technique is to be applicable to all valid inferences.

   b. Are you inclined to trust this procedure to establish the invalidity of the invalid argument in question 2? Explain why or why not.

## 5. Instantiation and Generalization

The distinction between universal and existential statements underlay both the trapezoid technique and the intuition exercised on bound variables in Section 4. In the trapezoid method, although each kind of statement was cast into trapezoids, each continued to be held apart from the other: universals were conjoined into a KU whereas existentials were one by one checked against that KU. Again in working with bound variables, the treatment given existentials was quite different from that given universals;

statements which, as far as their propositional content was concerned, were contradictory (*e.g.*, $Fx$ and $\overline{F}x$) were tolerated if (separately) quantified existentially. A similar contradiction between universally quantified statements, or between a universal and an existential one, constituted a real inconsistency.

In neither of these methods was any attempt made to remove the quantifying prefixes in order to rewrite an expression which the propositional logic alone could deal with. This reduction to propositional logic was in the trapezoid method, but the quantification determined the handling of the symbols. In bound variables, the quantifiers were always present to guide the intuition in confronting what was said to exist with what was said of everything.

It is now time to examine a means of stripping off the quantifiers to permit a completely unfettered use of the propositional logic. This time, accordingly, the very stripping off must incorporate some way of preserving the indispensable distinction. In short, the expression $Fx$ will supplant the expression $\exists x\, Fx$ or $x\, Fx$, but under restrictions and guides that will automatically respect the distinction. Four of these restrictions will be explained in this section; the remaining three, in the section to follow.

Consider this $Fx$, separated from any quantifier. Suppose $Fx$ to mean that $x$ is a farmer. Whether this statement is true or false plainly depends on who the $x$ is. If $j$ stands for Joe, then $Fj$ is a statement the truth of which depends on whether Joe is a farmer. $Fx$, stating that $x$ is a farmer, is false if $x$ is this textbook, false if $x$ is any of certain human beings, true if $x$ is any of certain other human beings.

Now if $Fx$ be derived from the universal statement that every $x$ is an $F$, $x\, Fx$, then its argument can represent any element in the universe, since all elements are declared to be $F$'s. To put it another way, if $x\, Fx$ be true, then $Fx$ follows, for $F$ must be true of $x$ *no matter which element x represents.* By contrast, $Fx$ can be made to follow from the existential assertion, $\exists x\, Fx$, only by adopting a special way of construing what the $x$ of $Fx$ is to refer to. The assertion that there exists an $F$ ($\exists x\, Fx$) can be made to imply that $x$ is a $F$ ($Fx$) provided that *by x we now mean one of those elements* (perhaps a unique element) *declared by $\exists x\, Fx$ to be an F.* When from the assertion $\exists x\, Fx$ we infer $Fx$, we thus have singled out some element in the universe and designated it by $x$. When from the assertion $x\, Fx$ we infer $Fx$, no such singling out has taken place; the $x$ of such an $Fx$ stands equally well for all elements.

Of course, one does not have to use $x$ to designate an element; $y$, $z$, or $w$, or any other letter (conventionally from the latter part of the alphabet) would serve as well. Consequently, from $\exists x\, Fx$ one can infer $Fy$, or $Fz$. Whatever letter is used represents the element declared by the implicans to be a farmer and thereby designates a particular element. Again, from $x\, Fx$ either $Fx$, $Fy$, or $Fz$, etc., can be inferred, but in this case the letter used still represents any element at all, not one in particular. This can be seen by carefully following the meaning of each line in this simple deduction:

There exists a farmer. All farmers vote. Therefore, there exists a voter.

1. $\exists x\ Fx$
2. $x(Fx \supset Vx)$ $\therefore \exists x\ Vx$
3. $Fy \supset Vy$ Inferred from line 2. The element here designated by $y$ can be any element in the universe. If it is not a farmer, line 3 is true by virtue of the falseness of the antecedent; if $y$ is a farmer, then it can safely be asserted that $y$ votes, since line 2 asserts that every farmer votes.
4. $Fy$ Inferred from line 1. Here $y$ now designates a particular element of the universe. It is true that $y$ is a farmer because by $y$ is here meant one of those elements spoken of in line 1.
5. $Vy$ 4,3 c
6. $\exists x\ Vx$ Inferred from line 5 by a reasoning to be discussed later.

Although as a general principle (illustrated by this deduction) the removal of the quantifier of $x\ Fx$ leaves the $y$ of $Fy$ capable of representing any element whatever, the interpretation of the lines within a proof may sometimes have to depart from this. Note the interpretation to be given the lines when the existential quantifier is removed before the universal one:

1. $\exists x\ Fx$
2. $x(Fx \supset Vx)$ $\therefore \exists x\ Vx$
3. $Fy$ Inferred from line 1. As in line 4 of the previous deduction, $y$ is here committed to designate one of the elements referred to in line 1.
4. $Fy \supset Vy$ Inferred from line 2. Because $y$ has already been made (by line 3) to refer to a farmer, it is not now the free-ranging representative of any element that it was in line 3 of the previous deduction. But however much $y$ is limited now, $Fy \supset Vy$ is nonetheless a valid inference. What line 2 asserts of every element is here invoked on behalf of the particular element designated in line 3 by $y$.

| | |
|---|---|
| 5. $Vy$ | 3,4 c |
| 6. $\exists x\ Vx$ | Inferred from line 5 by a reasoning to be discussed later. |

It can well be true of a universe that there exist elements that are farmers and other elements that are not: to assert both $\exists x\ Fx$ and $\exists x\ \overline{F}x$ constitutes no contradiction. Here is a fallacious stripping away of quantifiers leading to a spurious contradiction. The error is that, having resolved in line 3 to allow $y$ to designate a farmer, we have used it in line 4 to designate a non-farmer — patently a distinct element.

| | |
|---|---|
| 1. $\exists x\ Fx$ | Premise |
| 2. $\exists x\ \overline{F}x$ | Premise |
| 3. $Fy$ | Inferred from line 1. |
| 4. $\overline{F}y$ | (fallacious) Inferred from line 2. |
| 5. $Fy\overline{F}y$ | 3,4 R |

Here is a second deduction no less erroneous.

| | |
|---|---|
| 1. $\exists x\ Fx$ | Premise |
| 2. $\exists x\ Gx$ | Premise |
| 3. $Fy$ | From line 1. |
| 4. $Gy$ | (fallacious) From line 2. |

The error is less patent but otherwise no different. There is no assurance that the element declared in line 2 to be a $G$ is the same element that line 1 declares to be an $F$, yet line 4 makes precisely this assumption. From this can be seen the need for some kind of rule prohibiting such an assumption. It can be simply stated. Whenever $Fy$, for instance, is deduced from an existential statement, the representative resorted to (here the $y$) shall be a letter not formerly used in such a deduction.[14]

The moment has come for mastering some technical terminology. In a quantified expression, the symbolism lying within the scope of the quantifier

[14]The reader may wonder if any error can arise from this obligation to use a new letter. Suppose that the premises of this example really do refer to the same element — a unique element that is both an $F$ and a $G$.

| | |
|---|---|
| 1. $\exists x\ Fx$ | Premise |
| 2. $\exists x\ Gx$ | Premise |
| 3. $Fy$ | From line 1. |
| 4. $Gw$ | From line 2. |

Would the prescribed recourse to a letter in line 4 different from that in line 3 misrepresent the state of affairs? Not at all! All that has taken place is that the same element has been designated by two different letters. Recourse to $y$ and to $w$ does not assume that they are distinct elements; they might be two elements or they might be the same, and neither possibility is ruled out. On the other hand, it is impossible to strip away two quantifiers using the same letter in both cases without asserting thereby that the element is the same. In short, $y$ and $w$ might not be the same element or they might, whereas $y$ and $y$ are unavoidably the same.

is called its *matrix*.[15] The first $x$ in an expression such as $\exists x(Fx \vee Gx)$, *i.e.*, the $x$ in the quantifier, is said to be a *binding* occurrence; the second and third are *bound* ones. When $Fx \vee Gx$ is deduced from this statement, the two $x$'s are said to be *unbound* or *free* occurrences. What is more, the $x$ of $\exists x(Fx \vee Gx)$ is said to have been *instantiated* to itself; whereas if $y$ is chosen as an instance, or representative, *i.e.*, if $Fy \vee Gy$ be deduced from $\exists x(Fx \vee Gx)$, then the $x$ of the implicans is said to have been *instantiated* to $y$. The bound variable is frequently called the *instantiated* variable and the free one the *instantiating* variable. In this text the terms *variable* and *representative* will be preferred instead. The process of stripping away the quantifier by such a deduction is called *existential instantiation* (hereafter 'EI'); if the implicans be universally quantified, it is called *universal instantiation* ('UI'). The process of imposing quantifiers to again bind a representative, restoring it to the status of variable, is called *generalization*. The rule introduced in the previous paragraph can now be phrased in this language:

**Rule 1** *No two EIs within the same proof may be to the same representative.*

Here are two further examples of the kind of erroneous inference that is prevented by this rule. By ignoring its restriction, one can establish that because fish and fowl exist there must be something that is both:

| | | | |
|---|---|---|---|
| 1. $\exists x\, Hx$ | | 4. $Lx$ | 2 EI (erroneous) |
| 2. $\exists x\, Lx$ | $\therefore \exists x(HxLx)$ | 5. $HxLx$ | 3,4 R |
| 3. $Hx$ | 1 EI | 6. $\exists x(HxLx)$ | 5 EG (this step is explained below) |

or that because there are farmers, everything in the universe is a farmer:

| | | | |
|---|---|---|---|
| 1. $\exists x\, Fx$ | $\therefore x\, Fx$ | 4. $x\, Fx \vee \bar{F}x$ | 3 EI |
| 2. $x\, Fx \vee \sim x\, Fx$ | GP | 5. $Fx$ | 1 EI (erroneous) |
| 3. $x\, Fx \vee \exists x\, \bar{F}x$ | 2 TQ | 6. $x\, Fx$ | 5,4 c′ |

Before proceeding, it will be well to take note of a very important feature of instantiation, whether existential or universal. The representative chosen *must be used for each occurrence of the variable it replaces*. If from $x(Fx \supset Gx)$ one deduces $Fx \supset Gy$, this is less to be called an illicit instantiation than no instantiation at all! For the *raison d'être* of the arguments of the functions $F$ and $G$ in the implicans is to signify that the *same* element of the universe that is an $F$ is a $G$ (*cf.* Section 1 of this chapter). Accordingly, to let the $x$ of the matrix of $x(Fx \supset Gx)$, or of $\exists x(FxGx)$, become two possibly distinct

[15] In multiply quantified expressions the scope of the left-most quantifier includes those to its right; *e.g.*, in $x\, \exists y(Fx \equiv Gy)$—a simplified writing of $x(\exists y(Fx \equiv Gy))$, *cf.* p. 89, n12—$\exists y$ is within the scope of $x$. Accordingly the matrix of any quantifier consists of such functions as fall within the first parenthesis to its right (said parenthesis to be omissible if it would contain but one function, as in $x\, Fx$) and any quantifiers standing between itself and that parenthesis (or function).

elements of the universe by deducing $Fx \supset Gy$, or $FxGy$, is to destroy the utility of the entire symbolism. Naturally a statement such as $\exists xFx \cdot \exists xGx$ can (indeed, *must*, by the above rule) yield different representatives, but then each of the variables these representatives are replacing is bound by a distinct quantifier, so that there are two instantiations here instead of one.

(To deduce $x(FxGx)$ from $\exists y\, x(FxGy)$ would be equally mistaken. Far from instantiating the argument of $G$, such an operation illicitly places it under a different quantifier.)

To infer $x\ Fx$ from an implicans reading $Fx$ is to universally generalize (UG); to deduce $\exists x\ Fx$ from $Fx$ is to effect an existential generalization (EG). This operation, too, should be performed with care. Although the variable to which a representative is generalized (or by which it is quantified, to say the same thing differently) need not be the same letter as the representative, it must *not* be a letter that will place it *and another representative* under the same quantifier. To EG the representative $x$ in $Fxy$ to $\exists y\ Fyy$, for example, would be to do just this. Just as a *pseudo-instantiation* obscures the sameness of two occurrences of a variable in the implicans, so a *pseudo-generalization* obscures the distinctness of two representatives in the implicans.

It will later be seen that in both instantiations and generalizations the number of quantifiers may well exceed the number of distinct representatives. But *whenever the number of distinct representatives exceeds the number of distinct variables* in any pair of assertions, one of which is deduced from the other by generalization or instantiation, the operation is spurious. This provides an easy and practical way of identifying a pseudo-generalization or a pseudo-instantiation.

The most obvious and elementary rule governing UG is

> **Rule 2** *No representative to which an EI has been made may be UG'd.*

The soundness of this restriction becomes clear the moment the distinction between the EI and the UI is recalled. The representative in the $Fy$ derived from $\exists x\ Fx$ designates *some particular element* specified in $\exists x\ Fx$. To declare subsequently that what is true of $y$ is true of everything (*i.e.*, to deduce $x\ Fx$ from such an $Fy$) would be to set aside altogether the convention which justified the writing of $Fy$.

Here is a bald violation of this rule by which the world is peopled with farmers even more swiftly than in a previous fallacy:

| | |
|---|---|
| 1. $\exists x\ Fx$ | $\therefore\ x\ Fx$ |
| 2. $Fy$ | 1 EI |
| 3. $x\ Fx$ | 2 UG (erroneous) |

This breach throws light on generalization as an inference. Provided only that the assertion $Fy$ has been licitly written, its EG to $\exists x\ Fx$ is valid, for if no element were an $F$ how could $Fy$ have come to be justified? Thus, there is almost no restriction on the EG as an inference. A UG, on the other hand, makes the much broader assertion that everything has the property predicated of the representative. Such an inference is valid only if the representative indeed represents everything. An alternative way of expressing the above rule might be, *A UG may be effected only on representatives born of UIs.*[16] This inference shows a typical UG:

| | | | |
|---|---|---|---|
| 1. $x(Fx \supset Gx)$ | | 4. $Gx \supset Hx$ | 2 UI |
| 2. $x(Gx \supset Hx)$ | $\therefore x(Fx \supset Hx)$ | 5. $Fx \supset Hx$ | 3,4 c |
| 3. $Fx \supset Gx$ | 1 UI | 6. $x(Fx \supset Hx)$ | 5 UG |

There are various ways of making fallacious deductions by the misuse of UG. By using enough abandon, it is possible to prove that either everything is red or else nothing is:

| | |
|---|---|
| 1. $Rx \vee \overline{R}x$ | GP |
| 2. $x\, Rx \vee x\, \overline{R}x$ | 1 UG (erroneous) |

The rule blocking such fallacies is this:

**Rule 3** *Whenever any representative is generalized, the assertion in which it occurs must be freed of representatives (i.e., every representative within it must be somehow generalized).*[17] *Only one UG may be effected on any given representative therein.*

Here are some further deductions intended to clarify this rule. (Suppose line 3 in each case to be legitimately derived from previous lines and there to be no obstacle to UGing any of its representatives.)

| | |
|---|---|
| 3. $Gx \vee Hy \vee Jy$ | |
| 4. $Gx \vee \exists w(Hw \vee Jw)$ | 3 EG (erroneous; the representative $x$, occurring in the same assertion, has been left ungeneralized) |

| | |
|---|---|
| 3. $Fw \vee Gw \vee HzJz$ | |
| 4. $\exists x(Fx \vee Gx \vee .\, z\, Hz \cdot z\, Jz)$ | 3 EG UG (erroneous; the representative $z$ has been twice UG'd) |

[16]The reason the original statement of Rule 2 is to be preferred to this alternative involves the question of how Rule IV (GP) of the Deductive System is to be exploited in the quantificational logic. It is clearly desirable to keep any modification of the Deductive System to a minimum in accommodating the quantificational logic. This requires the avoidance of such GP's as $x(Fx \supset Fx)$ and $\exists x(Fx \vee \overline{F}x)$, tautologous though they be, because they are not in the specified forms (*cf.* p. 48). The more proper GP, $Fx \vee \overline{F}x$, could not be UG'd were this alternative expression of the rule adopted, whereas the rule as given permits a UG on such an expression provided $x$ has not elsewhere been the representative in an EI. GP's such as $x\, Fx \vee \sim x\, Fx$ are acceptable, of course, because they conform to one of the approved forms, namely, $p \vee \overline{p}$.

[17]The first part of this rule finds its justification in the next section. Its inclusion here simplifies the expression of this rule and a subsequent one.

Here are some examples complying with this rule:

3. $Gx \vee Hy \vee Jy$
4. $x(Gx \vee \exists w(Hw \vee Jw))$ — 3 UG EG

3. $Fw \vee Gw \vee HzJz$
4. $\exists x(Fx \vee Gx \vee z(HzJz))$ — 3 EG UG

3. $Gx \vee Hy \vee Jy$
4. $\exists x\, Gx \vee \exists x\, Hx \vee \exists x\, Jx$ — 3 EG (note that any representative (here $y$) may be EG'd any number of times)

3. $Fz\, Gz$
4. $x\, Fx \cdot Gz$ — 3 UG (note here that $Fz$ is itself asserted; each representative in that assertion has been generalized)

3. $Fz\, Gz$
4. $x\, Fx \cdot x\, Gx$ — 3 UG (see previous comment)

3. $Fw \vee Gw \vee HzJzLzMzNz$
4. $\exists x\, Lx \vee x\, Gx \vee .\ x(HxJx) \cdot \exists x\, Lx \cdot \exists x(MxNx)$ — 3 EG UG (the representative $w$ has been UG'd but once; this is also true of the representative $z$)

3. $Fw \vee Gw \vee HzJzLz$
4. $x(Fx \vee Gx) \vee x(HxJxLx)$ — 3 UG ($w$ is UG'd but once; $z$ is UG'd but once)

From these examples it is apparent that a representative need not be captured by a single generalization. Any number of generalizations can be used to capture it — a different generalization at each occurrence is permissible. The rule imposes two restrictions: (1) that only one such generalization on a given representative be a UG, and (2) that an assertion any part of which is generalized have no representatives at all remaining in it, *i.e.*, that it be entirely generalized.

Up to now, singulars have received no attention.[18] Singulars are never themselves quantified, to be sure, but what about the instantiation of other statements to the singular? If Charles is a burglar and all burglars are stealthy, it follows that Charles is stealthy. Such an argument is to be established by invoking the general statement on behalf of Charles in particular:

| | | | |
|---|---|---|---|
| 1. $Bc$ | | 3. $Bc \supset Sc$ | 2 UI |
| 2. $x(Bx \supset Sx)$ | $\therefore Sc$ | 4. $Sc$ | 1,3 c |

[18]It is conventional to use letters from the latter part of the alphabet for representatives and those from the early part for singulars. The application of the next rule respecting singulars requires that they be distinguishable from representatives.

It thus appears that a UI to a singular is altogether reasonable. Not so with an EI, lest it be provable that because someone is a farmer, Charles is:

| | |
|---|---|
| 1. $\exists x\, Fx$ | $\therefore Fc$ |
| 2. $Fc$ | 1 EI (erroneous) |

With respect to generalizations the situation is reversed. From the fact that some singular has a property it can licitly be deduced that something has that property (EG) but not that everything does (UG).

**Rule 4** *No EI may be made to, nor a UG effected on, a singular.*

These four rules extend the Deductive System developed in Chapter II to embrace the singly quantified monadic functional logic, *i.e.*, the parts of quantificational logic to which the previous two sections were dedicated. The next section provides rules by which the System becomes adequate for any problem within the quantificational logic so long as only the arguments of functions are subject to quantification.

Now that it is known what instantiations and generalizations are and what their uses are in making deductions, it is time to take up a matter which has been overlooked up to now. Under which of the four rules of the Deductive System are the deductions EI, UI, EG, and UG to fall? Are they reassertions, to be subtended under Rule III? Transformations under Rule I? Gratuitous Premises they clearly are not, so Rule IV can be excluded from consideration. Perhaps they are inferences, to be subtended under Rule II. If it be recalled that Rules I and III (unless aided by II) result in equivalent statements, the answer can be got by examining this simple deduction comprised exclusively of instantiation and generalization:

| | |
|---|---|
| 1. $x\, Fx$ | Premise |
| 2. $Fx$ | 1 UI |
| 3. $\exists x\, Fx$ | 2 EG |

Because the conclusion is not the equivalent of the premise but is instead implied by it, it is evident that one or both operations are inferences.

This is not a minor point! If they are inferences, then any implicans on which an instantiation or generalization is effected will have to qualify under Rule II as an expression on which an inference may be worked. This is to say that Rule II governs every instantiation and generalization over and above the specific rules here being laid down. The following deduction is fallacious not because any of the present rules are breached but because this general consideration has been overlooked.

| | | |
|---|---|---|
| If everything is quiet, the world has ended. Some things are quiet. So the world has ended. | 1. $x\, Qx \supset p$ | |
| | 2. $\exists x\, Qx$ | $\therefore p$ |
| | 3. $Qx$ | 2 EI |

| | |
|---|---|
| 4. $Qx \supset p$ | 1 UI (violates Rule II) |
| 5. $p$ | 3,4 c |

(Once a TH has been worked on line 1, so that the quantified expression is no longer under a horseshoe to its right, and a TQ has been worked on that quantified expression to remove its curl, the fallacy can only be brought off by a violation of Rule 1 of the present rules.) *If an expression does not qualify under Rule II, transformations designed so to qualify it must precede any instantiation or generalization.*

The schemata which these new inferences follow are these:

d. $x\ Fx$ implies $Fx$ (or $Fy$, $Fz$, $Fw$, etc.)

d′. $\exists x\ Fx$ implies $Fx$ (or $Fy$, $Fz$, $Fw$, etc.)

e. $Fx$ implies $x\ Fx$ (or $y\ Fy$, $z\ Fz$, $w\ Fw$, etc.)

e′. $Fx$ implies $\exists x\ Fx$ (or $\exists y\ Fy$, $\exists z\ Fz$, $\exists w\ Fw$, etc.)

It is of course understood that these schemata are to be used subject to the seven rules in this section and the next. The codes *UI*, *EI*, *UG*, and *EG* will be continued, but this should not be allowed to obscure the fact that these operations are a part of the Deductive System.[19]

As a convenient way of facilitating observance of the first two rules of this section, any solution involving an EI will be accompanied by an annotation in which is set down the *representative to which* each EI is effected. (The *variable* replaced by the representative is indifferent and need not be noted.) This will consist of 'EI' followed by each such EI'd representative. A subscript to each such letter can be used to indicate the line of the proof in which the EI appears. Compare the second and third examples below. In the section to come, this table will cease to be merely a convenience and become a necessity.

If anyone in the gang is in danger of discovery, they all are. Anyone in the gang in danger of discovery will shoot to kill when approached. Bill is in danger of discovery. So if he's in the gang, they will all shoot to kill when approached.

| | |
|---|---|
| $Gx$ | $x$ is a gang member |
| $Dx$ | $x$ is in danger of discovery |
| $Sx$ | $x$ will shoot to kill when approached |
| $b$ | Bill |

[19]With respect to these schemata only, quantifiers can be included among the operators allowed to stand over an expression qualifying under Rule II for an inference, for when permitted by the seven rules instantiation or generalization may be effected despite a higher ranking quantifier. For example, from $x\ y\ z\ Fxyz$ may be deduced $x\ y\ Fxyz$, and from this $x\ z\ y\ Fxyz$. Schemata of Rule I, of course, can be applied without these restrictions anyway (*cf.* the wording of Rule I).

| | | |
|---|---|---|
| 1. | $\exists x(GxDx) \supset x(Gx \supset Dx)$ | |
| 2. | $x(GxDx \supset Sx)$ | |
| 3. | $Db$ | $\therefore Gb \supset x(Gx \supset Sx)$ |
| 4. | $Gb \supset Gb$ | GP |
| 5. | $Gb \supset GbDb$ | 4,3 R |
| 6. | $Gb \supset \exists x(GxDx)$ | 5 EG |
| 7. | $Gb \supset x(Gx \supset Dx)$ | 6,1 c |
| 8. | $Gb \supset . \ Gx \supset Dx$ | 7 UI |
| 9. | $Gb \supset . \ \overline{G}x \vee Dx$ | 8 TH |
| 10. | $Gb \supset . \ \overline{G}x \vee GxDx$ | 9 TS |
| 11. | $GxDx \supset Sx$ | 2 UI |
| 12. | $Gb \supset . \ \overline{G}x \vee Sx$ | 10,11 c |
| 13. | $Gb \supset x(\overline{G}x \vee Sx)$ | 12 UG |
| 14. | $Gb \supset x(Gx \supset Sx)$ | 13 TH |

There is some fresh milk in the cupboard. Any milk is subject to souring if it's not refrigerated. Nothing in the cupboard is refrigerated. So something is subject to souring.

| | | | |
|---|---|---|---|
| $Fx$ | $x$ is fresh | $Rx$ | $x$ is refrigerated |
| $Mx$ | $x$ is milk | $Sx$ | $x$ is subject to souring |
| $Cx$ | $x$ is in the cupboard | | |

| | | | |
|---|---|---|---|
| 1. | $\exists x(FxMxCx)$ | | |
| 2. | $x(Mx \supset . \ \overline{R}x \supset Sx)$ | | |
| 3. | $x(Cx \supset \overline{R}x)$ | $\therefore \exists x \, Sx$ | |
| 4. | $My \supset . \ \overline{R}y \supset Sy$ | 2 UI | EI: $y_8$ |
| 5. | $My \supset . \ Ry \vee Sy$ | 4 TH | |
| 6. | $Cy \supset \overline{R}y$ | 3 UI | |
| 7. | $My \supset . \ \overline{C}y \vee Sy$ | 5,6 c″ | |
| 8. | $FyMyCy$ | 1 EI | |
| 9. | $\overline{C}y \vee Sy$ | 8,7 c | |
| 10. | $Sy$ | 8,9 c′ | |
| 11. | $\exists x \, Sx$ | 10 EG | |

In the above solution it may seem that after line 8 *My* (from line 8 by *b*) should be written before writing line 9. But this is omissible. The reference to line 8 in the justification of line 9 does not mean that every assertion from line 8 is here rewritten, but that the assertion here rewritten is to be found in line 8. The reference to 8 in the justification of line 10 is to be similarly construed. A line such as 4, or 9, constitutes only one assertion. If this is unclear, the definition of assertion (page 44) should be reviewed.

If the doctor is late, it's because someone got hurt in the game. But the only injuries were in the stands. No one in the stands was in the game. So the doctor is not late.

| | | | |
|---|---|---|---|
| $p$ | the doctor is late | $Gx$ | $x$ was in the game |
| $Hx$ | $x$ was hurt | $Sx$ | $x$ was in the stands |

| | | | | |
|---|---|---|---|---|
| 1. $p \supset \exists x(HxGx)$ | | | 6. $Hw \supset \overline{G}w$ | 5,4 c |
| 2. $x(Hx \supset Sx)$ | | | 7. $p \supset HwGw$ | 1 EI |
| 3. $x(Sx \supset \overline{G}x)$ | $\therefore \overline{p}$ | | 8. $\overline{p} \vee HwGw$ | 7 TH |
| 4. $Sw \supset \overline{G}w$ | 3 UI | EI: $w_7$ | 9. $\overline{p} \vee \overline{G}wGw$ | 8,6 c |
| 5. $Hw \supset Sw$ | 2 UI | | 10. $\overline{p}$ | 9 TS |

## The Deductive System Expanded for Quantificational Logic

| *The Original System* | *The Expansion* |
|---|---|
| I. *Transformations* (44) allow any expression to be replaced by its equivalent. | TQ (82): $x\ Fx \equiv \sim\exists x\ \overline{F}x$<br>$\sim x\ Fx \equiv \exists x\ \overline{F}x$ |
| TH, TE, TM, TD, TS | TQD (117)<br>TQS (118) |
| II. *Inference* (47) allows any expression over which stand only horseshoes to its left, dots, or wedges to be replaced by whatever could be deduced were it asserted. | UI (d) (100) $x\ Fx$ implies $Fx$ (or $Fy$, $Fz$, etc.)<br>EI (d′) $\exists x\ Fx$ implies $Fx$ (or $Fy$, $Fz$, etc.)<br>UG (e) $Fx$ implies $x\ Fx$ (or $y\ Fy$, $z\ Fz$, etc.) |
| a, b, c, c′, c″ | EG (e′) $Fx$ implies $\exists x\ Fx$ (or $\exists y\ Fy$, $\exists z\ Fz$, etc.)<br>(Inferences *d*, *d′*, *e*, and *e′* are regulated by the seven rules appearing on pages 113–114.)<br>f and f′ (154) |
| III. *Reassertion* (48) provides for conjoining or repeating previous assertions. *R*. | Quantified expressions can be rewritten with different variables under this rule; *e.g.*, $x\ \exists y\ Fxy$ can be rewritten as $w\ \exists y\ Fwy$ or as $z\ \exists w\ Fzw$. |
| IV. *Gratuitous Premises* (48) follow one of three forms: $p \vee \overline{p}$, $p \supset p$, or $p \equiv p$. *GP*. | A GP may not contain both a representative and a quantifier. (110) |

Numbers in parentheses refer to pages.
A more detailed chart appears at the back of the book.

## EXERCISE IV–5

1–5. Use the deductive technique on questions 2–a, 2–c, and 3 from page 91.

6. For each of the following arguments, write a pseudo-proof, setting down beside the erroneous line the number of the rule from this section which it violates.

   a. Alice is homesick. Some homesick persons put on weight. Therefore, Alice puts on weight.

   b. There are some homesick persons on the swimming team. Some homesick persons are glum. Hence, some swimming team members are glum.

   c. Some Juniors study late and are A students. Therefore, every Junior is an A student.

   d. All whales are ponderous. So there are no whales or else everything is ponderous.

7. What is fallacious about this deduction?

| | |
|---|---|
| 1. $Fx \equiv Fx$ | GP |
| 2. $x\ Fx \equiv \exists x\ Fx$ | 1 UG EG |

8. The instructor returned to McGinnis a deduction marked with X's as shown below. Despite considerable study, McGinnis could not determine what rules he had violated. To what passage in this section would you refer McGinnis to help him see his mistake?

| | | | |
|---|---|---|---|
| 1. $x(Kx \vee Lx\ .\!\supset Mx)$ | | | |
| 2. $\exists x(KxLx)$ | | | |
| 3. $x(Lx \supset Px)$ | $\therefore \exists x(KxMxPx)$ | | |
| 4. $Kx \vee Lx\ .\!\supset Mx$ | 1 UI | | EI: $x_8\ y_8$ |
| 5. $\bar{K}x\bar{L}x \vee Mx$ | 4 TH TM | | |
| 6. $\bar{K}x \vee Mx$ | 5 b | | |
| 7. $Ly \supset Py$ | 3 UI | | |
| 8. $KxLy$ | 2 EI | X | |
| 9. $Mx$ | 8,6 c′ | | |
| 10. $Py$ | 8,7 c | | |
| 11. $KxMxPy$ | 8,9,10 R | | |
| 12. $\exists x(KxMxPx)$ | 11 EG | X | |

9. Miss Bartlee had correctly arrived at the line $Fx \vee GyHy$ and could see no reason why her next line:

$$\exists x(Fx \vee \exists x(GxHx))$$

was marked wrong. How would you explain her error?

## 6. Shifting Variables

Because *shifting* will here mean *varying*, the term *shifting variable* may appear redundant until one takes a closer look at the meaning of *variable*. It will be recalled that in a statement such as $x(Fx \supset Gx)$ $x$ is both a place-holder signifying that the $x$ having the function $G$ is the same element having the function $F$, and also a convention — when so used as a quantifier — meaning that the matrix is true for each element in the universe. In this second way, $x$ can be said to be a variable; certainly not in the first way. In a statement such as $\exists x(FxGx)$ the place-holding use of $x$ is the same as before and the $x$ of the quantifier $\exists x$ means only that one or more elements exist of which this matrix is true; $x$ is here a variable only in the second use and not even there when the element referred to is unique. In the deduced $FyGy$ $y$ is called a representative rather than a variable because this no longer varies in any sense at all.

But there is a very important sense in which a variable can be said to shift. Taking $Oxy$ to mean that $x$ is an offspring of $y$ and supposing it is agreed that the universe contemplated consist only of human beings,[20] the statement that everyone is an offspring of someone or other is simply $x \exists y\, Oxy$. In such a statement the selection of the element $y$ plainly depends on the prior choice of the element $x$. The term *shifting variable* refers to such an existentially quantified variable, the choice of which depends on that of some universally quantified variable(s), necessarily of higher rank. That higher ranking universal, or rather the variable it binds, will be said to be a determinant of the shifting variable. For *shifting* the word *dependent* will often be used. It will make discussion easier to allow the terms *determinant* and *dependent* to be applied to the universal and existential quantifiers binding these variables as well as to the representatives to which they are instantiated.

Although $x$ in the expression $x \exists y\, Oxy$ is said to be a determinant of $y$, in $\exists y\, x\, Oxy$ no such determination is present. With the universal quantifier

[20]This is not to affirm that the universe *is* so limited, but only that as far as the *inferences to be entertained* are concerned we shall not introduce any element other than a human being. Such a limitation is often very convenient in undertaking a proof, and is warranted if all the elements involved in the argument have some feature in common (in this case, humanity). The final proof in the previous section was tacitly so limited, thus eliminating any need for the function $Px$ for *x is a person*.

thus subordinate to the existential, the choice of $y$ is to be made first, independently of the choice of $x$. This expression asserts the existence of some individual(s) — perhaps Kronos or Prometheus, if they were human, or Adam — from whom every individual is sprung.

The critical influence of these determinations on deductions in the quantificational logic is not hard to discern. If it be true that we are all offspring of some individual ($\exists y\ x\ Oxy$), then it follows that all are offspring of someone or other ($x\ \exists y\ Oxy$); but no implication exists in the opposite direction. For this reason the rules must provide for deducing the dependent statement from the independent while impeding the reverse deduction.

Most of the methods commonly in use provide this safeguard by a combination of four limitations, from which the method developed in this text will differ considerably. This paragraph and the three to follow will show how these four conventional rules provide the needed safeguards but in an excessively restrictive way. This examination will serve the double purpose of acquainting the student with these widely used rules and allowing him to see more clearly at what points the departures of this text are made. Of these four rules, then, the first is that only assertions can be instantiated or generalized. As has already been seen, this text's method allows parts of assertions — such as qualify under Rule II — to be instantiated or generalized. The second rule is that only one quantifier at a time, the major one (the outermost), can be stripped away or restored. The third is that an EI can never be made to any representative previously used either in an EI *or a UI*. The fourth is that no expression containing a representative born of an EI can be UG'd until that representative has first been generalized. By these rules the legitimate deduction of the previous paragraph would be:

| | |
|---|---|
| 1. $\exists y\ x\ Oxy$ | Premise |
| 2. $x\ Oxz$ | EI to the representative $z$ |
| 3. $Owz$ | UI to the representative $w$ |
| 4. $\exists y\ Owy$ | EG on the representative $z$ |
| 5. $x\ \exists y\ Oxy$ | UG on the representative $w$ |

and the outlawed deduction is rendered illicit by the combination of all four rules and particularly (in this example) by the fourth:

| | |
|---|---|
| 1. $x\ \exists y\ Oxy$ | Premise |
| 2. $\exists y\ Owy$ | UI to $w$ |
| 3. $Owz$ | EI to $z$ |
| 4. $x\ Oxz$ | UG on $w$ (illicit by conventional rules here contemplated because the implicans contains $z$ born of an EI) |
| 5. $\exists y\ x\ Oxy$ | EG on $z$ |

These provisions have the advantage of simplicity, but many a logician has cursed them because to keep the horse from being stolen they seem to have boarded up the barn door. Such broad control of dependent representatives unavoidably encumbers proofs in which no shifting variables are involved.

Consider the expression $x\ \exists y(FxGy)$. This consists essentially of two assertions — that everything is $F$ and that something is $G$, and there is no connection between the two except that they occur in a common matrix. The choice of $y$ certainly does not depend on that of $x$, and we easily intuit that the expression $\exists y\ x(FxGy)$ says exactly the same thing.

At this point, the solution may seem quite simple: if $y$ be quantified existentially under the universal quantification of $x$, the $x$ determines $y$ if they are arguments in some polyadic function; otherwise, *i.e.*, if they occur only in monadic functions, $y$ is independent of $x$. It is *almost* this simple, to be sure. But it is possible for purely monadic functions to set up a dependence. In $x\ \exists y(Fx \equiv Fy)$ $y$ is determined by $x$; only by allowing $x$ to be chosen first and then selecting an appropriate $y$ can the truth of this statement be defended. Notice that the statement is logically true; by choosing for $y$ the same element selected for $x$, the statement is always true no matter how the contemplated universe is constituted. The statement $\exists y\ x(Fx \equiv Fy)$, to which the first would be equivalent if no dependence were present, is quite false except in a universe in which everything is $F$ or in which nothing is.

One safe recourse is to allow for the possibility that any existential standing under a universal is dependent on it. But in following precisely this course, the conventional rules incur unfortunate complications. As an example, consider the following proof that $x\ \exists y(FxGy)$ implies $\exists y\ x(FxGy)$.[21]

| | | | |
|---|---|---|---|
| 1. | $x\ \exists y(FxGy) \quad \therefore\ \exists y\ x(FxGy)$ | | |
| 2. | $\exists y(FzGy)$ | 1 UI | EI: $w_3\ t_7$ |
| 3. | $FzGw$ | 2 EI | |
| 4. | $\exists y\ x(FxGy) \vee \sim\exists y\ x(FxGy)$ | GP | |
| 5. | $\exists y\ x(FxGy) \vee y\ \exists x(\overline{F}x \vee \overline{G}y)$ | 4 TQ | |
| 6. | $\exists y\ x(FxGy) \vee \exists x(\overline{F}x \vee \overline{G}w)$ | 5 UI | |
| 7. | $\exists y\ x(FxGy) \vee \overline{F}t \vee \overline{G}w$ | 6 EI | |
| 8. | $\exists y\ x(FxGy) \vee \overline{F}t$ | 3,7 c′ | |
| 9. | $Fz$ | 3 R | |
| 10. | $x\ Fx$ | 9 UG | |
| 11. | $Ft$ | 10 UI | |
| 12. | $\exists y\ x(FxGy)$ | 8,11 c′ | |

[21] Lines 6 and 7 of this deduction violate the first of the four common restrictions cited earlier, namely, that only assertions can be instantiated. Properly written, this conventional deduction would make a Provisional Assumption (*cf.* p. 53) of $\sim\exists y\ x(FxGy)$ for line 4, close this Assumption at line 9 with $\sim\exists y\ x(FxGy) \supset \overline{F}t$, and justify line 12 with 9, 11 $c''$. This illustrates why Provisional Assumptions are indispensable to most deductive systems.
This text's development of the rule obviates all four of these general restrictions.

To avoid such encumbering rules, the restriction must be confined to shifting variables. If line 4 of the illicit deduction on page 105 were regarded as illicit because line 3, the implicans, contains a *dependent*, the barn door could still be opened and closed at will. This will be the alternative pursued in this text.

> **Rule 5** *No UG may enter its quantifier under an existential one if the representative UG'd either determines that captured by the existential in question or is itself also EG'd.*[22]

This less restrictive rule allows the following straightforward deduction:

| | | |
|---|---|---|
| 1. $x\ \exists y(FxGy)$ | $\therefore \exists y\ x(FxGy)$ | $w$ |
| 2. $FwGz$ | 1 UI EI | EI: $z_2$ |
| 3. $xFx\ Gz$ | 2 UG | |
| 4. $FtGz$ | 3 UI | |
| 5. $\exists y\ x(FxGy)$ | 4 UG EG | |

The simultaneous instantiation of more than one quantifier (line 2), although constituting an economy permitted by the rules developed here, is not the point. The point is, rather, that the circuitous Provisional Assumption (or its equivalent) has been obviated. Notice that in the notations of the EI at the right, $w$ is placed over the letter $z$. This is to record that $z$ is dependent on $w$ as far as the technicalities are concerned (see below). Rule 5 would be violated were line 5 to be derived from line 2. The critical part of this proof is the generalizing and re-instantiating of the argument of $F$ so that in line 4 the representative to be UG'd has no dependent representative or, to say the same thing differently, in line 4 the $w$ on which $z$ depends is no longer present. In the previous proof this essential operation occurs in lines 9, 10, and 11. Here it occurs in lines 2, 3, and 4, manifesting a proper economy since this proof consists of virtually nothing else except this operation.

The notation in the table of EIs constitutes the apparatus of the present method for preventing fallacious proofs. The following fallacy merits comparison as to the parenthetical remarks with the same fallacy on page 105.

| | | |
|---|---|---|
| 1. $x\ \exists y\ Oxy$ | Premise | $w$ |
| 2. $Owz$ | 1 UI EI | EI: $z_2$ |
| 3. $\exists y\ x\ Oxy$ | 2 EG UG (Illicit by Rule 5, because $z$ depends on $w$ in the implicans) | |

If intuition is to be replaced by such record-keeping of dependences, some formal definition of dependence is needed to eliminate any uncertainty as to

[22]The last five words are occasioned by the need to outlaw deductions such as this:

| | |
|---|---|
| 1. $Pxy \vee \overline{P}xy$ | GP |
| 2. $\exists y\ x\ Pxy \vee \exists x\ y\ \overline{P}xy$ | 1 UG EG (violates Rule 5) |

whether or not a dependence is to be recorded.[23] Such a formal definition should identify the shifting variable as perfectly as possible. It has already been noted that if $y$ is existentially quantified under the universal quantification of $x$, there will be different $y$'s for different $x$'s if both are arguments of the same polyadic functions (as of $Oxy$). It was noted, too, that even monadic functions can be so arranged as to entail the shifting. Only one consideration remains. What if a second existential be tied to the first by still another function which requires it to shift along with the shifting of the first existential? For example, in

$$x\ \exists y\ \exists w(Oxy \cdot Oyw)$$

it is obvious that different choices of $x$ must result in corresponding shiftings in the selection of the $y$ whose offspring $x$ is, but what of the $w$ whose offspring $y$ is? Will it not also shift along with the shifting $y$ and therewith become a dependent of $x$?

The formal definition of what is to be construed as a dependent can now be adduced:

> One variable (call it $y$) will be construed as depending on another (call this one $x$) whenever:
>
> a. $y$ is existentially quantified within the scope of the universal quantification of $x$, and
>
> b. either $x$ itself or some variable dependent on $x$ occurs within the scope of $y$.

In $x\ \exists y\ \exists w(Oxy \cdot Oyw)$ the variable $y$ is dependent on $x$ because the *a* part of the definition is satisfied and so is the first condition under *b*. The variable $w$ is similarly dependent; indeed, it satisfies both conditions under *b*. Even if the scope of $\exists w$ were reduced — $x\ \exists y(Oxy\ \exists w\ Oyw)$ — it would still be dependent on $x$ by reason of the second condition under *b*, *i.e.*, in its scope $y$ occurs, and $y$ is a dependent on $x$.

Without doubt the reader is now wondering what is to be said of such expressions as $x\ \exists y(FxGy)$. If the formal definition purports to capture the essence of dependence, why does it also subsume some cases that are plainly not dependent? Simply because this is unavoidable. To be sure of including every instance of dependence, the definition unfortunately must include some instances of what may be called spurious dependences. Failure to subsume every genuine dependence would be intolerable, for such a weakness in the definition would permit a false deduction to be made. The inclusion of spurious dependences, on the other hand, is tolerable for two reasons. There are transformations which will allow any spurious dependence to be elim-

[23] When the student has completed the next section, to say nothing of Section 7 of Chapter V, any reluctance he may feel in saying goodbye to an intuitive procedure for identifying dependences will have been overcome.

inated from an expression. These will be set forth in the next section. And even if such transformations are not resorted to, every valid deduction can still be made despite the presence of a spurious dependence (provided it always falls under the definition). A spurious dependence may encumber the deduction but will not finally prevent it.[24]

From the necessity of keeping track of these dependences another requirement arises, *viz.*, that such inferences as may obscure them be barred. In making inferences from a line in which no dependence exists, no special care need be taken.

| | | |
|---|---|---|
| 1. $\exists y\ x\ Fxy$ | Premise | |
| 2. $x\ Fxy$ | 1 EI | EI: $y_2\ w_4$ |
| 3. $\exists y\ Fxy$ | 1 UI | |
| 4. $Fxw$ | 1 EI UI | |

Because the method of this text is oriented to the shifting variable, lines 3 and 4 of the above deduction are as legitimate as line 2. (Note that all are instantiations from line 1.) The forbidden inferences will be those that may lose track of a dependence. Consider a premise in which a shifting variable occurs.

| | | |
|---|---|---|
| 1. $x\ \exists y\ Fxy$ | Premise | |
| 2. $\exists y\ Fxy$ | 1 UI ?? | $x$ |
| 3. $x\ Fxy$ | 1 EI ?? | EI: $y_3\ w_4$ |
| 4. $Fxw$ | 1 EI UI | |

If from this line a line which instantiates *universally only* be deduced, some device will be needed for noting the dependence of a *variable* (the $y$ of line 2) on a *representative* (the $x$ of that line). An instantiation which is *only existential* will result in the dependence of a representative on a variable (line 3). The control of this will tend to be awkward. The alternative is simply to outlaw inferences like these in which the dependence is covert by requiring that determinant and dependent quantifiers be instantiated simultaneously as in line 4. By such simultaneous instantiations the dependence patent in line 1 in the *order of the quantifiers* is recorded for line 4 in the *table of EI's*.

> **Rule 6** *When one quantifier depends on another, instantiations must be made of both simultaneously or of neither.*

[24]If this were not so, the rules referred to earlier as commonly in use would render some valid deductions impossible instead of merely cumbersome; for by failing to define dependence explicitly the way the present Deductive System does, they implicitly classify every existential quantifier as dependent on every universal standing over it no matter what the matrix of the existential contains.

Out of this caution respecting covert dependences it will be well to examine this expression:

$$\exists y\, Fxy \vee y\, \overline{F}xy$$

Despite its tautologous appearance and its resemblance to the schema $p \vee \overline{p}$, such a statement is inadmissible as a GP because it conceals a dependence.

| | | |
|---|---|---|
| 1. $\exists y\, Fxy \vee y\, \overline{F}xy$ | GP (erroneous) | |
| 2. $Fxw \vee y\, \overline{F}xy$ | 1 EI | EI: $w_2$ |
| 3. $\exists w\, x(Fxw \vee y\, \overline{F}xy)$ | 2 EG UG | |

For suppose $Fxy$ to mean *x is father to y;* not only will the second disjunct be the one that renders the statement true for some human beings (the non-fathers) whereas the first disjunct renders it true for fathers, but the existentially quantified $y$ of the first disjunct must shift with the various fathers. By ignoring this dependence, one can deduce the far-from-tautologous assertion that there is someone with respect to whom everyone is either *his* father or nobody's (line 3).[25] Now no restriction governing GPs can be a part of these seven rules controlling instantiation and generalization. Instead of a gloss on Rule II (Inferences), it will be related to Rule IV of the Deductive System (see page 49). But the very reasons that occasion Rule 6 call for this limitation as well:

*A GP may not contain both a representative and a quantifier.*

Although the end of the matter of the shifting variable is getting near (only one rule is yet to be adduced), a critical complication is still to be considered. In the following deduction the error is plain enough to the intuition. From the premise that everyone is an offspring of someone or other the conclusion is deduced that someone is his own offspring.

| | | |
|---|---|---|
| 1. $x\, \exists y\, Oxy$ | Premise | $z$ |
| 2. $Ozz$ | 1 UI EI ?? | EI: $z_2$ |
| 3. $\exists x\, Oxx$ | 2 EG | |

The fallacy does not lie in the transition from line 2 to line 3, for if $z$ bears the relation $O$ to himself then clearly such an element exists in the universe: the third line is unquestionably deducible from the second. The instantiations in the second line, though violating no rule so far developed, contain the fallacy.

The simplest resolution of this difficulty would be to prohibit a dependent variable from being instantiated to the same representative as its determinant.

[25]Line 1 can be licitly deduced from certain non-tautologous lines, of course. It can even be written as a line in a null-premise deduction, but in that case $x$ is derived from an EI and the present conclusion, along with every other non-tautology, is barred.

But this would provide only for the above case, which is but a special instance of the fallacy. For consider this argument:

| | | |
|---|---|---|
| 1. $x\ \exists y\ Oxy$ | Premise | $w \quad z$ |
| 2. $Owz$ | 1 UI EI | EI: $z_2\ w_3$ |
| 3. $Ozw$ | 1 UI EI ?? | |
| 4. $OwzOzw$ | 2,3 R | |
| 5. $\exists x\ \exists y(OxyOyx)$ | 4 EG | |

Here another fallacious conclusion casts suspicion on only one line — line 3. The error here is that having stated in line 2 "Let $w$ be $O$ to $z$", line 3 proposes that that same $z$ shall be $O$ to that same $w$. That $z$ should be offspring to some element or other is correct enough (the premise asserts this of every element in the universe), but that the element to which it is $O$ should be precisely the same element that is $O$ to it — this is not necessarily the case at all.

The generalizing of this fallacy can more easily be understood from yet a third argument of the same order:

| | | |
|---|---|---|
| 1. $x\ \exists y\ Oxy$ | Premise | |
| 2. $Oxy$ | 1 UI EI | $x \quad y \quad z$ |
| 3. $Oyz$ | 1 UI EI | EI: $y_2\ z_3\ x_4$ |
| 4. $Ozx$ | 1 UI EI ?? | |
| 5. $OxyOyzOzx$ | 2,3,4 R | |
| 6. $\exists x\ \exists y\ \exists z(OxyOyzOzx)$ | 5 EG | |

Lines 2 and 3 are free from error. From them one could licitly deduce

| | |
|---|---|
| 4′ $OxyOyz$ | 2,3 R |
| 5′ $\exists x\ \exists y\ \exists z(OxyOyz)$ | 4′ EG |

which concludes that there is something $O$ to something which is in turn $O$ to something. The fourth line of the fallacious argument sets it astray, again with the same error as before — a kind of illicit circuit-closing. This is what the last rule must preclude.

Once again the annotations in the EI table are of great help. When the dependences recorded there are carefully examined, the circuit-closing common to each of the above examples can be discovered. Up to line 3 and including it, the deduction is free from error; the instantiations are licit.

$$\begin{array}{c} x \quad y \\ \text{EI: } y_2\ z_3 \end{array}$$

But together these two instantiations have made the choice of $z$ dependent on that of $x$. For in line 3 $z$ is determined by $y$, and line 2 establishes that the choice of $y$ depends in turn on that of $x$. Due to this combination of dependences, $x$ determines not only $y$ but through it $z$ as well. Now if these additional determinations are recorded in the table parenthetically, *they can also be mechanically controlled there.* If $x$, which already stands over the first mention of $y$ in this control, is now written above the second occurrence of $y$ when $y$ becomes a determinant, the determination of $z$ by $x$ will be recorded.

$$\begin{array}{ll} & \quad (x) \\ & x \quad y \\ \text{EI:} & y_2 \; z_3 \end{array}$$

If the same procedure is pursued when the erroneous line 4 is attempted, *i.e.*, if each determinant of $z$ is entered again over the appearance of $z$ as a determinant, it will be evident that $x$ is determining itself (or, if you prefer, is dependent on itself) and this is the very fallacy that characterized the first inference in this series.

$$\begin{array}{ll} & \qquad\quad (x) \\ & \quad (x)(y) \\ & x \quad y \quad z \\ \text{EI:} & y_2 \; z_3 \; x_4 \end{array}$$

In short, if whenever a representative is both a dependent and a determinant its own determinants are entered above it *at each occurrence*, the undesirable circuit-closing can be ruled out in all its forms quite simply:

> **Rule 7** *No instantiation by which any representative becomes its own determinant is licit (even when that determination derives from other instantiations).*

The following illustration of the use of this control in a more complex series of instantiations will repay scrutiny:

1. $\exists x\ y\ z\ \exists w\ t\ Dxyzwt \qquad \therefore\ y\ \exists w\ z\ \exists t\ \exists x\ Dxyzwt$
2. $Dxyzwt$ — 1 UI EI
3. $C^{26} \vee \exists y\ w\ \exists z\ t\ x\ \overline{D}xyzwt$ — GP TQ
4. $C \vee \overline{D}xyzwt$ — 3 UI EI (violates Rule 7)

$$\begin{array}{lcccc} & & & & (y) \\ & & & y & (z) \\ & & & z & w \\ \text{EI:} & x_2 & w_2 & y_4 & z_4 \end{array}$$

Note that $t$, instantiated in line 2, makes no appearance in the table. This is because it is a UI having no dependent. Both $t$ and $x$ are missing from the notations for line 4 for the same reason. The $x$ (line 2) and $y$ (line 4) are entered in the table without letters above them because they are EIs of

[26] 'C' is simply an abbreviated way of writing the conclusion.

independent variables. The table records two determinations from line 2 since $w$ is dependent on both $y$ and $z$. Line 4 is disallowed by the rule because the dependence of $z$ on $w$, once the specified bookkeeping enters the determinants of $w$, entails $z$'s determination of itself.

Below this paragraph the seven rules adduced in this and the previous section are gathered together. To their right are listed the restrictions commonly in use. The two listings are made to correspond as nearly as possible. The advantages of the rules adopted here are principally these: (1) deductions are considerably disembarrassed by (a) permitting instantiations and generalizations to be made on *parts* of lines instead of entire lines only, and (b) by largely freeing non-shifting variables from controls required only for shifting ones; (2) these rules, where they differ from conventional ones, go more directly and less artificially to the heart of the fallacies that beset instantiation and generalization. It follows that the student will not only be enabled to write proofs more direct and economical than those required under the methods replaced here, but he can also expect to master more thoroughly this critical part of quantificational logic. In this text, the present set of rules has a third *raison d'être*. The theoretical aspects of the cross-outs treated later in this chapter will be more readily understood by means of these rules.

| Rules Governing Instantiation and Generalization | Restrictions Typical of Other Methods |
|---|---|
| 1. No two EIs may be made to the same representative (95). | 1. No EI may be made to a representative used before in any instantiation, whether an EI or a UI. |
| 2. No representative to which an EI has been made may be UG'd (96). | 2. Same |
| 3. Whenever any representative is generalized, the assertion in which it occurs must be freed of representatives. Only one UG may be effected on any given representative therein (97). | 3. Only entire lines are generalized, so that every occurrence of any representative is necessarily bound. It is permissible to leave other representatives uncaptured. |
| 4. No EI may be made to, nor a UG effected on, a singular (99). | 4. Same |
| 5. No UG may enter its quantifier under an existential one if the representative UG'd | 5. No line containing a representative to which an EI has been made may be UG'd. |

either determines that captured by the existential in question or is itself also EG'd (107).

6. When one quantifier depends on another, instantiations must be made of both simultaneously or of neither (109).

7. No instantiation by which any representative becomes its own determinant is licit (even when that determination derives from other instantiations) (112).

6. Quantifiers are removed (always the highest ranking) one at a time and restored in the same way.

7. Because all existential quantifiers standing under universals are handled as if they involved dependences, the previous rules suffice for this control.

### EXERCISE IV–6–A

1. Here are eight attempts to deduce $\exists y\ x\ Gxy$ from $x\ \exists y\ Gxy$. In each case identify the erroneous line and the rule it violates.

a.

| | | | |
|---|---|---|---|
| 1. | $x\ \exists y\ Gxy$ | Premise | EI: $z_2$ |
| 2. | $x\ Gxz$ | 1 EI | |
| 3. | $\exists y\ x\ Gxy$ | 2 EG | |

b.

| | | | |
|---|---|---|---|
| 1. | $x\ \exists y\ Gxy$ | Premise | |
| 2. | $Gwz$ | 1 UI EI | $w$ |
| 3. | $x\ Gxz$ | 2 UG | EI: $z_2$ |
| 4. | $\exists y\ x\ Gxy$ | 3 EG | |

c.

| | | | |
|---|---|---|---|
| 1. | $x\ \exists y\ Gxy$ | Premise | |
| 2. | $Gwz$ | 1 UI EI | $w$ |
| 3. | $\exists y\ Gwy$ | 2 EG | EI: $z_2\ s_4$ |
| 4. | $Gws$ | 3 EI | |
| 5. | $\exists y\ x\ Gxy$ | 4 UG EG | |

d.

| | | | |
|---|---|---|---|
| 1. | $x\ \exists y\ Gxy$ | Premise | |
| 2. | $\exists y\ Gwy$ | 1 UI | |
| 3. | $\exists y\ x\ Gxy$ | 2 UG | |

e. 1. $x\,\exists y\, Gxy$ — Premise
   2. $x\, Gxy \vee \exists x\, \overline{G}xy$ — GP
   3. $x\, Gxy \vee \overline{G}xy$ — 2 EI
   4. $Gxy$ — 1 UI EI
   5. $x\, Gxy$ — 4,3 c′
   6. $\exists y\, x\, Gxy$ — 5 EG

   $x$
   EI: $x_3\; y_4$

f. 1. $x\,\exists y\, Gxy$ — Premise
   2. $Gst$ — 1 UI EI
   3. $\exists y\, x\, Gxy$ — 2 UG EG

   $s$
   EI: $t_2$

g. 1. $x\,\exists y\, Gxy$ — Premise
   2. $Gxy \vee \overline{G}xy$ — GP
   3. $\exists y\, x\, Gxy \vee \exists x\, y\, \overline{G}xy$ — 2 UG EG
   4. $\sim\exists x\, y\, \overline{G}xy$ — 1 TQ
   5. $\exists y\, x\, Gxy$ — 4,3 c′

h. 1. $x\,\exists y\, Gxy$ — Premise
   2. $\exists y\, Gwy \vee y\, \overline{G}wy$ — GP
   3. $\exists y\, Gwy \vee \overline{G}wy$ — 2 UI
   4. $Gwy$ — 1 UI EI
   5. $\exists y\, Gwy$ — 4,3 c′
   6. $\exists y\, x\, Gxy$ — 5 UG

   $w$
   EI: $y_4$

2. Each of the following proofs that is completed is fallacious. For these, justify each line, complete any EI table, identify the fallacious line and its error. Where no proof is given, write a valid one (always tabulating any EI's, of course).

a. 1. $\exists y\, x\, Lyx \quad \therefore\; x\,\exists y\, Lxy$
   2. $Lzw$
   3. $x\,\exists y\, Lxy$

b. 1. $\exists y\, x\, Lxy \quad \therefore\; x\,\exists y\, Lxy$

c. 1. $x\, Gxx \quad \therefore\; x\, y\, Gxy$
   2. $Gww$
   3. $x\, y\, Gxy$

d. 1. $x\, Gxx \quad \therefore\; x\, y\, Gxy$
   2. $Gwz$
   3. $x\, y\, Gxy$

e. 1. $x\, Gxx \quad \therefore\; x\,\exists y\, Gxy$

f. 1. $x\, Hxxx \quad \therefore\; x\,\exists y\, Hxyx$

g. To prove: $\exists y\, x(\overline{H}x \vee Hy)$
   1. $\overline{H}w \vee Hw$
   2. $\exists y\, x(\overline{H}x \vee Hy)$

h. To prove: $\exists y\, x(\overline{H}x \vee Hy)$

i. To prove: $\exists y\, x(Gx \equiv Gy)$
   1. $Gx \equiv Gx$
   2. $\exists y\, x(Gx \equiv Gy)$

j. To prove: $x\,\exists y(Gx \equiv Gy)$

k. 1. $x\ \exists y(Fxy \vee Hy)$
   2. $y\ \exists x\ \overline{F}xy \quad \therefore \exists x\ Hx$
   3. $Fwz \vee Hz$
   4. $\overline{F}wz$
   5. $Hz$
   6. $\exists x\ Hx$

l. 1. $x\ Kx \supset x\ Lx$
   2. $\exists x\ Kx \quad \therefore x\ Lx$
   3. $Kx$
   4. $Kx \supset Ly$
   5. $Ly$
   6. $x\ Lx$

m. 1. $x\ Kx \supset x\ Lx$
   2. $\exists x\ Kx \quad \therefore x\ Lx$
   3. $\exists x\ \overline{K}x \vee x\ Lx$
   4. $\overline{K}x \vee x\ Lx$
   5. $Kx$
   6. $x\ Lx$

n. 1. $x\ \exists y(Gy \vee Mxy) \quad \therefore x\ y\ \overline{M}xy \supset x\ Gx$
   2. $Gy \vee Mxy$
   3. $x\ y\ \overline{M}xy \supset x\ y\ \overline{M}xy$
   4. $x\ y\ \overline{M}xy \supset \overline{M}xy$
   5. $x\ y\ \overline{M}xy \supset Gy$
   6. $x\ y\ \overline{M}xy \supset x\ Gx$

o. 1. $x\ \exists y(Gy \vee Mxy) \quad \therefore x\ y\ \overline{M}xy \supset \exists x\ Gx$

p. 1. $x\ y(Tx \supset Sy) \quad \therefore y(\exists x\ Tx \supset Sy)$
   2. $Tx \supset Sy$
   3. $y(\exists x\ Tx \supset Sy)$

q. 1. $x\ y(Tx \supset Sy) \quad \therefore y(\exists x\ Tx \supset Sy)$

## EXERCISE IV–6–B

1. Given the premise $x\ \exists y\ z\ \exists w\ Fxyzw$, some of the following conclusions are valid and some are not. Write deductions for each of them, noting the violated rule (and where it is violated) in any fallacious proof.

   *a. $\exists x\ \exists y\ \exists z\ Fxyyz$
   b. $\exists x\ \exists y\ \exists z\ Fxyxz$
   c. $x\ \exists y\ \exists z\ Fxyxz$
   d. $\exists x\ \exists y\ Fxyyx$
   e. $\exists x\ \exists y\ Fxyxy$
   f. $\exists x\ \exists y\ \exists z\ Fxyzx$

2. Prove each of these:

   a. 1. $x(Cx \supset Fx) \quad \therefore x\ y(CxDyx \supset \exists z(FzDyz))$
   b. 1. $x\ \exists y\ Fxy \quad \therefore x\ \exists y\ \exists z(FxyFyz)$
   c. 1. $x\ \exists y\ Fxy \quad \therefore x\ \exists y\ \exists z(FxyFzy)$

3. Only one of these is valid. Prove that one.

   a. 1. $x\ \exists y\ w(Fxyw \vee Gxw)$
      2. $x\ \exists y\ \overline{G}yx \quad \therefore \exists y\ w\ \exists x\ Fxyw$

   b. 1. $x\ \exists y\ w(Fxyw \vee Gxw)$
      2. $x\ \exists y\ \overline{G}yx \quad \therefore \exists y\ \exists w\ \exists x\ Fxyw$

   c. 1. $x\ \exists y\ w(Fxyw \vee Gxw)$
      2. $x\ \exists y\ \overline{G}xy \quad \therefore \exists y\ w\ \exists x\ Fxyw$

## 7. The Scope of Quantifiers

Certain equivalences which can be proved without much difficulty with the apparatus already constituting the Deductive System will throw light on the previous section and make the ensuing section much easier. Because they are useful, thesc two new transformations are incorporated into the System.

One of them is properly called Quantifier Distribution. Whatever fits either of these schemata can be justified as *TQD*:

$$x\,Fx \cdot x\,Gx \mathrel{.\!\equiv} x(FxGx)$$
$$\exists x\,Fx \vee \exists x\,Gx \mathrel{.\!\equiv} \exists x(Fx \vee Gx) \qquad \text{TQD}$$

It will be noted that the universal quantifier can be collected or distributed over the dot; the existential, over the wedge.

Certain operations might appear to conform with these schemata when they do not. For example,

$$x\,\exists y\,Fxy \vee \exists y\,Gy \tag{1}$$

cannot be transformed by TQD to

$$\exists y(x\,Fxy \vee Gy)$$

What the schema in question authorizes is the collection of quantifiers standing over their respective disjuncts. The last quantifier of (1) stands over that disjunct, but the first existential quantifier is not the major operator of the disjunct in which it stands. Even

$$\exists y\,Fy \vee \exists x\,\exists y\,Gxy$$

would have to be changed to

$$\exists y\,Fy \vee \exists y\,\exists x\,Gxy$$

before it could be TQD'd to

$$\exists y(Fy \vee \exists x\,Gxy)$$

A transformation by TQD can be said to change the scope of a quantifier — incidentally, at least; but its principal effect is to change the *number* of quantifiers. A second transformation, TQS, has the change of the *scope* of quantifier(s) as its sole effect. TQS will designate such transformations as the following:

$x(Fx \vee y\,Gy)$ is equivalent to
$x\,y(Fx \vee Gy)$ which in turn is equivalent to
$y\,x(Fx \vee Gy)$ or to
$x\,Fx \vee y\,Gy$

Instead of offering schemata for this transformation (the above are not schemata), it is more convenient to lay down certain conditions each of which any licit TQS must satisfy:

Two expressions differing only in the scope of some quantifier, $Q$,[27] are equivalent provided that

1. in both expressions $Q$ binds exactly the same arguments, and

| *Licit* | *Illicit* |
|---|---|
| 1. $x\ \exists y(FxGxy)$ | 1. $x\ \exists y(FyGxy)$ |
| 2. $x(Fx\ \exists y\ Gxy)$ 1 TQS | 2. $x(Fy\ \exists y\ Gxy)$ (wrong) |

2. no unlike quantifier standing over $Q$ in one expression stands under it in the other, unless they are independent in both expressions,[28] and

| *Licit* | *Illicit* |
|---|---|
| 1. $\exists x\ \exists y\ z\ Fxyz$ | 1. $\exists x\ \exists y\ z\ Fxyz$ |
| 2. $\exists y\ \exists x\ z\ Fxyz$ 1 TQS | 2. $\exists x\ z\ \exists y\ Fxyz$ (wrong) |
| ($\exists x$ and $\exists y$ are *like* quantifiers) | |
| | 1. $x\ \exists y(FxGy)$ |
| | 2. $\exists y\ x(FxGy)$ (wrong) |
| 1. $w\ x(\exists y\ Axwy \lor \exists y\ \exists z\ Bzwy)$ | |
| 2. $w\ \exists z\ x(\exists y\ Axwy \lor \exists y\ Bzwy)$ | 1. $x\ \exists y\ w(Axyw \lor \exists z\ Bzw)$ |
| 1 TQS | 2. $\exists z\ x\ \exists y\ w(Axyw \lor Bzw)$ |
| (in both expressions $\exists z$ and $x$ are independent) | (wrong; $\exists z$ and $w$ are not independent in both lines) |

3. no negation, equivalence, or horseshoe-to-its-right standing over $Q$ in one expression stands under it in the other.

| *Licit* | *Illicit* |
|---|---|
| 1. $x(Fx \supset \exists y\ Gxy) \supset p$ | 1. $x\ Fx \supset \exists y\ Gy$ |
| 2. $x\ \exists y(Fx \supset Gxy) \supset p$ 1 TQS | 2. $x(Fx \supset \exists y\ Gy)$ (wrong) |
| 1. $x\ Fx \equiv .\ y\ Gy \lor \exists w\ Hw$ | 1. $x\ Fx \equiv .\ y\ Gy \lor \exists w\ Hw$ |
| 2. $x\ Fx \equiv .\ y\ \exists w(Gy \lor Hw)$ 1 TQS | 2. $x(Fx \equiv .\ y\ Gy \lor \exists w\ Hw)$ (wrong) |
| 1. $\sim\exists x(Px \lor \exists y\ Qy)$ | 1. $\sim\exists x(Px \lor \exists y\ Qy)$ |
| 2. $\sim\exists y\ \exists x(Px \lor Qy)$ 1 TQS | 2. $\exists y \sim\exists x(Px \lor Qy)$ (wrong) |

TQS can be very useful in ridding an expression of a dependence that is apparent only.

1. $x\ \exists y(FxGy)$
2. $x(Fx\ \exists y\ Gy)$ 1 TQS
3. $\exists y\ x(FxGy)$ 2 TQS

[27]Reference to a single quantifier, $Q$, simplifies discourse here. However many TQSs may be effected simultaneously, the three conditions are to be satisfied with respect to each quantifier TQS'd; *e.g.*, $x(\exists yFxy \cdot \exists zGxz)$ is licitly TQS'd to $x\ \exists y\ \exists z(FxyGxz)$, and $x\ \exists y(FxGy)$ to $x\ Fx\ \exists y\ Gy$.

[28]A universal quantifier and an existential quantifier are unlike in the sense intended here. For the examples under this heading the student will do well to keep in mind the definition of dependence given on p. 108.

Note that the premise of this deduction must be construed as involving a dependence. The transformations demonstrating that the dependence is not a genuine one use a strategy that can be called typical. Line 1 shows dependence because the existential quantifier is under the universal and its matrix contains both variables. In line 2 the dependence has disappeared because within the scope of the existential, $Gy$, $x$ no longer occurs. Note that the second condition under TQS has been observed because the existential quantifier still stands under the universal. Line 3 places the existential over the universal but still without violence to the second condition because in both line 2 and line 3 the variables $x$ and $y$ are independent. By thus reducing the scope of the existential until the universally bound variable is excluded, such spurious dependences can be eliminated. TQD can also be very helpful in these shortenings of matrices, as these more complex examples show:

1. $w\ x\ \exists y(AxPy \lor Bw\overline{P}y \lor \overline{A}x\overline{B}w)$ $\therefore$ $w\ \exists y\ x(AxPy \lor Bw\overline{P}y \lor \overline{A}x\overline{B}w)$
2. $w\ x(\exists y(Bw\overline{P}y) \lor \exists y(AxPy \lor \overline{A}x\overline{B}w))$ — 1 TQD
3. $w\ x(\exists y(Bw\overline{P}y) \lor Ax\ \exists y\ Py \lor \overline{A}x\overline{B}w)$ — 2 TQS
4. $w(\exists y(Bw\overline{P}y) \lor x(Ax\ \exists y\ Py \lor \overline{A}x\overline{B}w))$ — 3 TQS
5. $w(\exists y(Bw\overline{P}y) \lor \exists y\ x(AxPy \lor \overline{A}x\overline{B}w))$ — 4 TQS
6. $w\ \exists y(Bw\overline{P}y \lor x(AxPy \lor \overline{A}x\overline{B}w))$ — 5 TQD
7. $w\ \exists y\ x(AxPy \lor Bw\overline{P}y \lor \overline{A}x\overline{B}w)$ — 6 TQS

1. $\exists w\ z\ \exists x\ y(AxwyBxCzwy)$ $\therefore$ $\exists w\ \exists x\ y\ z(AxwyBxCzwy)$
2. $\exists w\ z\ \exists x(y(AxwyBx)\ y\ Czwy)$ — 1 TQD
3. $\exists w\ z(\exists x\ y(AxwyBx)\ y\ Czwy)$ — 2 TQS
4. $\exists w\ \exists x\ z(y(AxwyBx)\ y\ Czwy)$ — 3 TQS
5. $\exists w\ \exists x\ z\ y(AxwyBxCzwy)$ — 4 TQD
6. $\exists w\ \exists x\ y\ z(AxwyBxCzwy)$ — 5 TQS

A final remark about TQD has been deferred until now simply because its elucidation is easier once TQS is understood. Despite the fact that TQD is a transformation, one limitation must be made. Recall that even Rule IV required a restriction (that no GP mingle representatives and variables) lest a dependence be obscured. The same danger can occur when the matrices of quantifiers collected by TQD contain representatives. The fallacious deduction shown here illustrates this.

1. $x\ \exists y(FxGy \lor \overline{F}x\overline{G}y)$ — Premise
2. $x(\exists y(FxGy) \lor \exists y(\overline{F}x\overline{G}y))$ — 1 TQD
3. $x(Fx\ \exists yGy \lor \overline{F}x\ \exists y\overline{G}y)$ — 2 TQS
4. $Fx\ \exists yGy \lor \overline{F}x\ \exists y\overline{G}y$ — 3 UI
5. $\exists y(FxGy) \lor \exists y(\overline{F}x\overline{G}y)$ — 4 TQS
6. $\exists y(FxGy \lor \overline{F}x\overline{G}y)$ — 5 TQD (wrong)
7. $\exists y\ x(FxGy \lor \overline{F}x\overline{G}y)$ — 6 UG

The dependence of $y$ on $x$ which was properly destroyed by lines 2 and 3 (which made licit the ensuing UI without a simultaneous EI) is restored in line 6 *and there obscured.*

For this reason:

> *A TQD collecting two quantifiers in both of whose matrices the same representative occurs is illicit.*

**EXERCISE IV–7–A**

1. Without recourse to TQD, write deductive proofs establishing each of its two schemata.
2. With respect to $x\, Fx \vee x\, Gx$ and $x(Fx \vee Gx)$, does each imply the other, one imply the other, or neither imply the other? Write a proof for any implication you believe exists.
3. Which of the following derivations are proper? If any are illicit, state why.

a.

| | | |
|---|---|---|
| 1. | $\exists x\, Fx \vee \exists y\, Gy \vee \exists z\, w\, Hwz$ | Premise |
| 2. | $\exists x\, Fx \vee \exists x\, Gx \vee \exists x\, w\, Hwx$ | 1 R |
| 3. | $\exists x(Fx \vee Gx \vee w\, Hwx)$ | 2 TQD |

b.

| | | |
|---|---|---|
| 1. | $\exists x\, Fx \cdot \exists x\, Gx$ | Premise |
| 2. | $\exists x(FxGx)$ | 1 TQD |

c.

| | | |
|---|---|---|
| 1. | $x(Fx \cdot \exists y\, Gyx)$ | Premise |
| 2. | $x\, Fx \cdot x\, \exists y\, Gyx$ | 1 TQD |
| 3. | $x\, Fx \cdot y\, \exists x\, Gxy$ | 2 R |

d.

| | | |
|---|---|---|
| 1. | $x\, Fx \cdot \exists y\, x\, Gxy \cdot y\, x\, Hxy$ | Premise |
| 2. | $x\, Fx \cdot \exists y\, Gxy \cdot x\, Hxy$ | 1 UI |
| 3. | $x\, Fx \cdot x\, \exists y\, Gxy \cdot x\, y\, Hxy$ | 2 UG |
| 4. | $x(Fx \cdot \exists y\, Gxy \cdot y\, Hxy)$ | 3 TQD |

e.

| | | | |
|---|---|---|---|
| 1. | $\exists z\, x(\exists y\, Axyz \vee \exists y\, Bxyz)$ | Premise | |
| 2. | $x(\exists y\, Axyz \vee \exists y\, Bxyz)$ | 1 EI | EI: $z_2$ |
| 3. | $x\, \exists y(Axyz \vee Bxyz)$ | 2 TQD | |

f.

| | | | |
|---|---|---|---|
| 1. | $\exists z\, x(\exists y\, Axyz \vee \exists y\, Bxyz)$ | Premise | |
| 2. | $\exists z\, x\, \exists y(Axyz \vee Bxyz)$ | 1 TQD | EI: $z_3$ |
| 3. | $x\, \exists y(Axyz \vee Bxyz)$ | 2 EI | |

4. Which of the following are proper TQSs of the expression $y\ \exists x(Fxy \vee Gx \vee Hy)$? If any are illicit, state why.
   a. $y\ \exists x(Gx \vee Hy \vee Fxy)$ c. $y(Hy \vee \exists x(Gx \vee Fxy))$
   b. $y(Gx \vee Hy \vee \exists x\, Fxy)$ d. $\exists x(Gx \vee y(Hy \vee Fxy))$

5. Prove these. *N.B.*: Do not effect a TQS and a TQD at the same time; this may yield an illicit line. Remember that apart from instantiation or generalization any change in the *number* of quantifiers requires a TQD.
   a. 1. $x\ \exists y(Fx \vee Gy)$ $\therefore\ \exists y\ x(Fx \vee Gy)$
   b. 1. $x\ \exists y(Fy \cdot Gx \vee Hy)$ $\therefore\ \exists y\ x(Fy \cdot Gx \vee Hy)$
   c. 1. $x\ \exists y(Fx \equiv \overline{F}y)$ $\therefore\ \exists y\ \exists w\ x(Fx\overline{F}y \vee \overline{F}xFw)$
   d. 1. $w\ x\ \exists y(AxPy \vee Bw\overline{P}y)$ $\therefore\ \exists y\ x\ w(AxPy \vee Bw\overline{P}y)$

6. Complete the following proof. Notice that deduced lines appeal to Rule I. (Why?)

   To prove: $x\ \exists y(Fx \equiv Gy) \equiv .\ \exists xFx \supset \exists yGy \cdot \exists x\overline{F}x \supset \exists y\overline{G}y$

   1. $\exists xFx \supset \exists yGy \cdot \exists x\overline{F}x \supset \exists y\overline{G}y$
   $.\equiv \exists xFx \supset \exists yGy \cdot \exists x\overline{F}x \supset \exists y\overline{G}y$ GP
   2. $x\overline{F}x \vee \exists yGy \cdot xFx \vee \exists y\overline{G}y$
   $.\equiv \exists xFx \supset \exists yGy \cdot \exists x\overline{F}x \supset \exists y\overline{G}y$ 1 TH TQ
   3. $x(\overline{F}x \vee \exists yGy) \cdot x(Fx \vee \exists y\overline{G}y)$
   $.\equiv \exists xFx \supset \exists yGy \cdot \exists x\overline{F}x \supset \exists y\overline{G}y$ 2 TQS
   4. $x(\overline{F}x \vee \exists yGy \cdot Fx \vee \exists y\overline{G}y)$
   $.\equiv \exists xFx \supset \exists yGy \cdot \exists x\overline{F}x \supset \exists y\overline{G}y$ 3 TQD
   5.
   $.\equiv \exists xFx \supset \exists yGy \cdot \exists x\overline{F}x \supset \exists y\overline{G}y$ 4 TD
   6.
   $.\equiv \exists xFx \supset \exists yGy \cdot \exists x\overline{F}x \supset \exists y\overline{G}y$ 5 TS
   7.
   $.\equiv \exists xFx \supset \exists yGy \cdot \exists x\overline{F}x \supset \exists y\overline{G}y$ 6 TQS
   8.
   $.\equiv \exists xFx \supset \exists yGy \cdot \exists x\overline{F}x \supset \exists y\overline{G}y$ 7 TQD
   9. $x\ \exists y(Fx \equiv Gy) \equiv .\ \exists xFx \supset \exists yGy \cdot \exists x\overline{F}x \supset \exists y\overline{G}y$ 8 TE

7. a. Prove this argument:

   1. $x\ \exists y(Cx \supset Py \cdot Px \supset Cy)$ $\therefore\ \exists x\, Cx \supset \exists y\, Py \cdot \exists x\, Px \supset \exists yCy$

   (Hint: This proof is easy if line 2 is instantiations.)

   b. The premise is what Tim wrote to symbolize the statement "If any book on this shelf is cloth-bound, there is also some paperback on this shelf and if any paperback is on the shelf there must be some cloth-bound book there, too." The conclusion is what Tom wrote for the same statement. To see their equivalence places no small demand on the intuition. Another quite rigorous challenge is to prove equivalence, whether by adding to the above proof another which uses the conclusion as premise and vice versa, or by going from either expression to the other by transformations only.

### EXERCISE IV–7–B

Prove these arguments:

1. If Charles is afraid of anything, it's either a thunderstorm or a debutante. Aunt Susan is no debutante. So, provided she's not a thunderstorm, Charles isn't afraid of her.

   $Axy$ $x$ is afraid of $y$ $c$ Charles $a$ Aunt Susan
   $Tx$ $x$ is a thunderstorm $Dx$ $x$ is a debutante

2. 1. $x\,\exists y(Axy \vee Gy)$
   2. $\exists x\, y\, \overline{A}xy$ $\therefore \exists x\, Gx$

3. 1. $\exists y(Hy\overline{G}y)$ $\therefore \exists y\, x(\overline{F}xHy \vee Fx\overline{G}y)$

4. Alice and at least one of her friends are soccer fans. All soccer fans know some Spanish. Hence, at least one of Alice's friends knows some Spanish and so does Alice.

   $a$ Alice $Fxy$ $x$ is a friend of $y$
   $Sx$ $x$ is a soccer fan $Kx$ $x$ knows some Spanish

5.*a. Which rule prevents you from proving that either something is prior to everything or there is something to which nothing is prior?
   b. Prove that either something is prior to everything or everything has something or other that it isn't prior to.
   c. It is false that something is prior to itself. So it is false that something is prior to everything.

6. 1. $x\,\exists y(FxGy \vee Hx\overline{G}y)$
   2. $x(Fx \equiv Hx)$ $\therefore \exists y\, x(FxGy \vee Hx\overline{G}y)$

*7. There are some students whom every teacher regards as scholars. Adams and Bennett are teachers. Therefore, there is some student whom they both regard as a scholar. (Among the functions, let $Rxy$ mean *x regards y as a scholar.*)

8. 1. $\exists x\, y(\overline{K}yx\, Mx)$
   2. $x(Mx \supset y(Kxy \vee Cxy))$
   3. $\exists y\, x(\overline{C}xy \cdot \exists w\, \overline{K}wy)$ $\therefore \exists x(Mx\, Cxx) \cdot \exists x\, \overline{M}x$

9. 1. $x\, Fx$ $\therefore x\, \exists y\, w(AxBy \vee \overline{B}y \vee \overline{A}xBw)$

*10. a. To prove: $x(\exists y\, Fxy \vee y\, \overline{F}xy)$
   b. To prove: $\exists x(\exists y\, Fxy \vee y\, \overline{F}xy)$

## 8. Translating from Language to Symbols

Any natural language such as English is redundant, ambiguous, and idiomatic, which three features respectively lessen the difficulty of communication, increase it, and conceal it. When translating from so rich but uncertain a language as ours into the rigorously precise language of logical symbolism, the problem is to determine the meaning the English intends and capture it in the new language. For the first part of this task, the reader's life-long acquaintance with English must be adequate. To help him with the second part is the purpose of this section.

The student is urged to discard any hope of relying mechanically on words such as *some*, *any*, *all*, or even *and*! A short list of plain usages of these will reveal how unplain they would become if interpreted literally.

If something goes wrong, Al frets. $\exists x\, Wx \supset Fa$

This offers no difficulty, but notice that in English we can make the same statement by wording the antecedent *If anything goes wrong*. It is not the wording but the meaning that must be captured. A nearly similar statement can require very different symbolizing.

If something goes wrong, Al frets about it. $x(Wx \supset Fax)$

This, too, could as well begin *If anything*, but the meaning with either beginning can be captured only by resort to a universal, $x(Wx \supset Fax)$, since the consequent contains a reference to an argument occurring in the antecedent. For if, misled by the use of the word *something*, one were to write $\exists x(Wx \supset Fax)$, the result would be entirely unsatisfactory. (This will be considered more fully below.) Nor do all the times we use the word *all* merit the universal, as this sentence attests. When *all* is coupled with a negative, it is frequently equivalent to a universal prefixed by a curl, *i.e.*, to an existential statement. And often this coupling is overlooked if a mechanical approach is taken. By a literal interpretation *All Buicks are not expensive* would appear to be an E-form statement, when its real content is an O-form one — that some Buicks are not expensive. As for *and*,

All the men and women are sick. $x(Mx \vee Wx \,.\!\supset Sx)$

would be disastrously misconstrued by the symbolism $x(MxWx \supset Sx)$, which asserts that whatever is both a man and a woman is sick. Even when *and* occurs in the predicate it can sometimes require a wedge:

All that Frances buys for the
department is gloves and hats. $x(Bfx \supset\!. \; Gx \vee Hx)$

Colloquial speech not only abounds in these freewheeling usages of superficially precise words but is also rich in treacherous idiomatic constructions. *Only nations are members of the UN* is an A-form statement, to be sure, yet it scarcely means that all nations are members of the UN. *All except the A's are B* is a construction which in some contexts may mean both that those things not $A$'s are $B$ and those things that are $A$'s are not $B$; whereas in other contexts it should be limited to the first of these meanings.

Meaning obviously depends on wording. But in view of these examples the first advice for translating from English into the language of logic is

1. Be guided by the meaning rather than by the wording of the English statement.

Each time a proposition is symbolized a decision must be made as to how much detail to incorporate into the functions, *i.e.*, what functions to use. This depends on the inference being analyzed.

> If the court gave any of the orders we requested, no funds can legally be removed from the DA's safe. But if no money can be legally removed from his safe, his political ruin is certain. Therefore, if the court gave any of the orders we requested, the DA's political ruin is certain.

This argument requires no quantification whatever, but can be symbolized quite adequately in the propositional logic as

$$1.\ p \supset q$$
$$2.\ q \supset r \qquad \therefore p \supset r$$

But contrast this argument with it:

> If the court gave any of the orders we requested, no funds can legally be removed from the DA's safe. But funds are being legally removed from his safe. So if the court gave an order, it is not the one we requested.

A modicum of quantification is now demanded to express properly this inference

$Cx$ $x$ is an order given by the court
$Rx$ $x$ is an order we requested
$p$ Funds are legally being removed, etc.

$$1.\ \exists x(RxCx) \supset \overline{p}$$
$$2.\ p \qquad \therefore x(Cx \supset \overline{R}x)$$

A further variation of the argument will evoke further quantification.

> If the court gave any of the orders we requested, no funds can legally be removed from the DA's safe. But the court has given some orders and funds are nevertheless being removed from the DA's safe. Therefore, either the orders given were not any that we requested or the removal of funds is illegal.

Besides the previous functions, something like these will be required:

$Fx$ $x$ is a removal of funds from the DA's safe
$Lx$ $x$ is legal

$$1.\ \exists x(RxCx) \supset x(Fx \supset \overline{L}x)$$
$$2.\ \exists x\, Cx \cdot \exists x\, Fx \qquad \therefore x(Cx \supset \overline{R}x) \vee \exists x(Fx\overline{L}x)$$

Notice that the *nevertheless*, like the *but* in the same premise, is simply discarded as irrelevant verbiage. The conclusion also merits attention. Were the first disjunct to be written as an O-form, $\exists x(Cx\overline{R}x)$, the conclusion would still be provable, but it would no longer be the conclusion given. For $\exists x(Cx\overline{R}x)$ states that the court gave orders not requested by us, whereas the conclusion given requires a complete separation of orders given from those requested, which is only to be expressed with the E-form.

2. In determining what functions to use, be guided by what detail seems to be necessary for capturing the essence of the reasoning.

Without other comment a third advice can conveniently be set down at this point:

3. Develop the ability to read back into English exactly what any given symbolism says. This will enable you to see if what you have written in symbols corresponds to what you intended to write.

The remainder of this section will consider in more detail some of the characteristics of the language into which the propositions are cast — the symbolism. As the complexity of the proposition grows, that of the symbolism grows, too. As this takes place, the translator will find the four Aristotelian forms more and more useful. For apart from the exceptions later to be noted, every quantification, with its matrix, will constitute one of the standard forms. This is true whether the matrix corresponds to the whole of some quite complex statement or merely to one of its constituent parts.

Indian tribes of the American West in the nineteenth century were invariably suspicious of the agreements and promises of the white man's governmental or military authorities.

Suppose that the reasoning being analyzed requires these functions:

| | |
|---|---|
| $Ix$ | $x$ is an Indian tribe in the nineteenth century |
| $Wx$ | $x$ is of the West |
| $Sxy$ | $x$ is suspicious of $y$ |
| $Ax$ | $x$ is an agreement |
| $Px$ | $x$ is a promise |
| $Mxy$ | $x$ makes $y$ |
| $Gx$ | $x$ is a governmental authority of the white man |
| $Nx$ | $x$ is a military authority of the white man[29] |

To illustrate the usefulness of a continued attention to the Aristotelian forms, let us undertake this translation by starting with the most general parts of the proposition and proceeding from there to more and more detail. This process will not be unlike that of parsing a sentence. The subject is *Indian*

[29] $Mx$ would serve; the other $M$ entailed being a dyadic function, there need be no confusion between the two. The use of another letter, however, may help avoid even needless confusion.

*tribes*, variously qualified, and the predicate is *were suspicious of agreements and promises*, these, too, with modifying clauses. The sense of *invariably* is that all such Indian tribes were suspicious. (Even without this word, this interpretation would be a likely one.) Hence this informal mixture of language and symbols:

*x*(*Ix* — qualified somehow ⊃ *Sx*? — the '?' to be elaborated)

The qualification of the subject is not difficult:

$$x(IxWx \supset Sx?)$$

but the elaboration of the predicate is more troublesome. The question mark can be replaced by *y* easily enough, and the things to be remarked further with respect to *y* can be noted:

*x*(*IxWx* ⊃ (Quantifier) (*Sxy*, *y* being agreements or promises made by a governmental authority of the white man or a military authority of the white man))

What form this inner parenthesis is to take may well be the next question. The choice is between an I-form — there are some agreements or promises, etc., and the *x* (the Indian tribe) is suspicious of them; and the A-form — every such agreement or promise is the object of such suspicion. The latter is meant, plainly. So now, within the main matrix, an A-form appears:

$$x(IxWx \supset y( \quad \supset \quad ))$$

Any perplexity as to whether *Sxy* is subject or predicate of this A-form is quickly resolved. If it be subject, then something must subsequently be predicated about the *y*; if it be predicate, then it is being predicated about some *y* determined in another way. That is, if we write

*y*(*Sxy* ⊃ *y* is an agreement or promise, etc.)

we shall be stating that everything the Indian tribe is suspicious of is an agreement or promise, etc.; whereas by writing

*y*(*y* is an agreement or promise, etc. ⊃ *Sxy*)

we shall be stating that every *y* so qualified is the object of the tribe's suspicion. Clearly this last is required.

*x*(*IxWx* ⊃ *y*(*y* is an agreement or promise of a governmental authority of the white man or a military authority of the white man ⊃ *Sxy*))

Recourse to the functions for the informally expressed portion leads to:

$x(IxWx \supset y(Ay \vee Py$ and $y$ was made by a governmental, etc. $.\supset Sxy))$

The dot by the horseshoe is called for by the fact that in an A-form the horseshoe must necessarily be major. From the sense it is also plain that the conjunction contained in the written word 'and' following the disjunction must stand over that disjunction.

$x(IxWx \supset y(Ay \vee Py \cdot M?y$ — '?' being a governmental, etc. $.\supset Sxy))$

How and where is the $z$ to replace '?' to be quantified? An A-form written thus

$$z(Gz \vee Nz .\supset Mzy)$$

declares that everything that is either a governmental or a military authority of the white man is a maker of $y$. All this is hardly required for the tribe's suspicions to arise. Rather it is enough that some such authority be the maker of $y$. It will not do merely to change the quantifier and leave the horseshoe — for this is to depart egregiously from a standard form. A straightforward I-form is indicated as the proper condition for the tribe's suspicion of $y$:

$$\exists z(Gz \vee Nz \cdot Mzy)$$

Therefrom, the completed symbolism:

$$x(IxWx \supset y(Ay \vee Py \cdot \exists z(Gz \vee Nz \cdot Mzy) .\supset Sxy))$$

Scrutiny of the punctuation — always a fitting precaution — shows that all is in order. The major horseshoe is the first (despite the dot by the last one) since the entire inner A-form (the matrix quantified by $y$) is but the predicate of the over-all A-form. The inmost quantification, the I-form, is a conjunction as it should be. It is also a conjunct in the subject of the inner A-form, so that the punctuation of operators within the one can be ignored in examining the ranking of operators in the other.

In conformity to the third advice, it will be well to read back into English the symbolism now completed. The syntax of the reading can afford to be somewhat battered — it must be if we are to follow the order of the quantifications and operators — but the *sense of the original* must be there if the symbolism is correct.

> Take what you will, if it be an Indian tribe of the nineteenth century and of the West then whatever is either an agreement or a promise and there exists some governmental authority of the white man or some military authority of his that made this agreement or promise aforementioned — whatever is such an agreement or promise will be the object of $x$'s suspicion.

The fourth advice:

4a. With certain exceptions (to be noted below), it is well to be guided in each quantification by one of the Aristotelian forms.

In the second paragraph of this section, "If something goes wrong, Al frets about it" was properly written $x(Wx \supset Fax)$, with the remark that $\exists x(Wx \supset Fax)$ would be entirely unsatisfactory. The latter is a departure from the standard forms, being neither an A-form, because of the quantifier, nor an I-form, because of the implication. To read it back into English is to discover its weakness. It says that there exists something which, if it goes wrong, is fretted about by Al. This statement is unquestionably true. Its defect is precisely that it is *too easily* true — vacuously true. For notice that all that is needed to make the implication true is a false antecedent, so that this statement merely alleges the existence of something of which either the antecedent is false or the consequent is true. If Al never had a thing to worry about, this statement would still be true simply by offering one's right thumb, or the square root of two, or anything else that doesn't go wrong as the element of the universe whose existence the statement alleges. Notice that this objection cannot be raised against $x(Wx \supset Fax)$; the universal quantifier is a challenge to pick what element we will in the universe. The allegation is that the matrix will be true for every one of them. Accordingly, it is not only true for one's right thumb and the square root of two, but for whatever might really go wrong. This is a *bona fide* rendition of the original instead of a spurious, emptily-true one.

By the same token, the matrix of an I-form or an O-form is rarely to be quantified universally. The results are likely to state much more than is intended; $x(FxGx)$ asserts that everything in the universe is an $F$ and a $G$.

The *components* of standard forms may of course be disjunctions or conjunctions, or any other propositional constants. Examples of disjunctions and conjunctions in the antecedent or the consequent of A-forms have already appeared several times. The equivalence is by no means barred. To state that any good scholar will make Dean's List if and only if he applies himself requires that the antecedent — corresponding always to the subject of a universal — be the *good student* whereas what is predicated of him is that his making the Dean's List is equivalent to his applying himself:

$$x(Gx \supset. Dx \equiv Ax)$$

In this way an implication can justifiably be one of the conjuncts in an I-form:

Someone set fire to this barn
and if he's in there he's dead. $\exists x(Px \cdot Sxb \cdot Ixb \supset Dx)$

The matrix of this statement is a conjunction. The mistaken rendition of "If something goes wrong, Al frets about it," $\exists x(Wx \supset Fax)$, departs from the standard forms in that an implication, rather than a conjunction, is major under an existential quantifier.

This is a good place to warn the student against one particular form. For "Every weekend Alice dates someone" it is correct to write:

$$x(Wx \supset \exists y(Py\ Dayx)) \qquad (1)$$

or, what is the same thing by TQS:

$$x\ \exists y(Wx \supset Py\ Dayx) \qquad (2)$$

But "Everyone seeing a ghost is mystified" provides an easy opportunity to err badly. This writing is correct:

$$x(Px\ \exists y(Gy\ Sxy) \supset Mx) \qquad (3)$$

or (anticipating what the next advice will make clearer) this writing is also correct:

$$x\ y(Px\ Gy\ Sxy \supset Mx) \qquad (4)$$

but it is quite wrong to write:

$$x\ \exists y(Px\ Gy\ Sxy \supset Mx) \qquad (5)$$

for in this latter (erroneous) symbolism an *existential* stands *over* an implication in the *antecedent of which is a variable bound thereby*. Notice that in (2) the existential stands over the implication, but that nothing it binds occurs in the antecedent. In (3) the existential, although binding a variable in the antecedent, stands *under* the implication. A writing on the order of (5) is *almost always* a mistaken writing.

Having seen these distinctions, the reader is now prepared for an explanation of why an apparently counterfeit form may after all be correct. When every element considered in an inference has some common characteristic, it may be convenient to limit the universe considered to that kind of element only. For instance, when we resolved in Section 6 to consider only human beings in connection with the function $Oxy$, $x$ is an offspring of $y$, we were dealing with such a limited universe of discourse. Examine the argument (from page 91) about Bill when expressed in two alternative ways, the one on the right being its symbolizing in a limited universe:

Bill is a person who farms. Any farmer likes to eat. Hence, there is someone who likes to eat.

| | |
|---|---|
| 1. $PbFb$ | 1. $Fb$ |
| 2. $x(Fx \supset Lx)$ $\quad\therefore \exists x(PxLx)$ | 2. $x(Fx \supset Lx)$ $\quad\therefore \exists x\ Lx$ |

Now when such a limitation is exploited, an I-form without one of the conjuncts it would otherwise have may occur (*cf.* the above conclusions). Thus it is possible for a statement such as "Someone either has a key or knows how to open doors without a key" to be written

$\exists x(Hx \vee Ox)$

instead of

$\exists x(Px \cdot Hx \vee Ox)$

The first form, when TH'd, would resemble exactly the form condemned above. Only by considering it as an ellipsis of the writing on the right does it seem to conform to the I-form. So, in a limited universe, might the statement that every man is a virtuous sinner result in $x(VxSx)$, an elliptical version of $x(Px \supset VxSx)$.

There can even be correct writings of this sort without any limitation of the universe: correct writings of false statements (that everything is pink: $x\,Px$) or of true statements that really do apply to every element in the universe (that everything is identical with itself: $x\,Ixx$). But the tenor of the next advice is by now obvious:

> 4b. Unless some limitation of the universe justifies it, exercise extreme caution in using a universal *without* a horseshoe as the major operator in its matrix and in using an existential *with* such a major operator.

From his study of the scope of quantifiers, the student will recall that the situation of quantifiers in an expression is sometimes indifferent. For example,

$$x(Fx \supset \exists y(Gy\; z(Hz \supset Jxyz))) \quad \text{and} \quad x\; \exists y\; z(Fx \supset .\; Gy \cdot Hz \supset Jxyz)$$

are equivalent expressions. That on the left incorporates the Aristotelian forms more clearly and may for this reason be preferred; but a writing resembling that on the right may sometimes suggest itself to the intuition, in which case seeing the equivalence between the two statements is the key to checking out the writing as to Aristotelian forms.

Being familiar with two other equivalences is a great boon to symbolizing. Because

$$p \supset .\; q \supset r \quad \text{is equivalent to} \quad pq \supset r$$

the student who is pondering a choice between these two:

$$x(AxBx \supset:\; Cx \vee Dx\; .\!\supset Fx) \qquad x(AxBx \cdot Cx \vee Dx\; .\!\supset Fx)$$

is wasting his time. The second equivalence is restricted to expressions *in the consequent of which no argument occurring in the antecedent recurs.* In these,

$$x(Px \supset p) \quad \text{is equivalent to} \quad \exists x\, Px \supset p$$

so that two expressions such as

$$x(Fx \vee Gx\; .\!\supset \exists y\; \exists z(HyJzMyz))$$

and

$$\exists x(Fx \vee Gx) \supset \exists y\; \exists z(HyJzMyz)$$

express the same proposition. Symbolisms (3) and (4) in the discussion preceding advice 4–b embody this same equivalence.

> 5. Be familiar with the equivalences just cited.

## Guides for Translating from Language to Symbols

1. Be guided by the meaning rather than by the wording of the English statement.
2. In determining what functions to use, be guided by what detail appears to be necessary for capturing the essence of the reasoning.
3. Develop the ability to read back into English exactly what any given symbolism says. This will enable you to see if what you have written in symbols corresponds to what you intended to write.
4. a. With certain exceptions, it is well to be guided in each quantification by one of the Aristotelian forms.

   b. Unless some limitation of the universe justifies it, exercise extreme caution in using a universal *without* a horseshoe as the major operator in its matrix and in using an existential *with* such a major operator.
5. Be familiar with TQS and the two equivalences cited on page 130.

### EXERCISE IV–8–A

1. Read each statement back into colloquial English.

| | | | |
|---|---|---|---|
| $Ax$ | $x$ is an apple | $Ex$ | $x$ is an elephant |
| $Bx$ | $x$ is bitter | $Fxy$ | $x$ is friendly to $y$ |
| $Cxy$ | $x$ chases $y$ | $Gx$ | $x$ is a goat |
| $Dx$ | $x$ is a dog | | |

a. $x(Ax \supset Bx)$

b. $\exists x\, \exists y(Dx\, Ay\, Cxy)$

c. $\exists x(Dx\, y(Ey \supset Cxy))$

d. $\exists x(Ex\, y(Dy \supset Cyx))$

e. $\exists x\, \exists y(Ex\, Gy\, Cxy\, Fxy)$

f. $\exists x\, \exists y(Ex\, Cxy\, Fxy)$

g. $x(Gx \supset y(Fyx \supset Cxy))$

h. $x\, y(Gx\, Fyx \supset Cxy)$

i. $\exists x\, y(Gx \cdot Fxy \supset Cyx)$

j. $\exists x(Gx\, Bx\, y\, \overline{F}xy)$

k. $x(Dx \supset \exists y\, Cxy)$

l. $x\, \exists y(Dx \supset Cxy)$

m. $\exists y\, x(Dx \supset Cxy)$

n. $\sim\exists x(DxGx)$

o. $x(DxGx \supset \sim\exists y\, Cxy)$

p. $x\, y(DxGx \supset \overline{C}xy)$

q. $x\, y(Dx \vee Ax\, . \supset \overline{C}yx)$

r. $x(Dx \supset y\, z(GyByCyz \supset Fxz))$

s. $x\, y\, z(DxGyByCyz \supset Fxz)$

t. $x\, z(Dx\, \exists y(GyByCyz) \supset Fxz)$

*2. Read these statements back into colloquial English.

| | | | |
|---|---|---|---|
| $f$ | Fred | $Px$ | $x$ is a person |
| $Lxyz$ | $x$ is an occasion on which $y$ looks at $z$ | $Sxy$ | $x$ is an occasion on which $y$ shaves |
| $Mx$ | $x$ is a morning | | |

a. $x(Mx \supset Sxf)$

b. $\exists x(MxSxf)$

c. $x\ y(Px\ Lyxf \supset \overline{S}yf)$

d. $y(\exists x(Px\ Lyxf) \supset \overline{S}yf)$

e. $\exists x(Px\ y(Syx \supset \overline{L}yxx))$

f. $x(Mx\ y(Py \supset \overline{L}xyf) \supset \overline{S}xf)$

g. $x\ y(PxPy \supset \exists z(LzxyLzyx))$

h. $x\ y\ \exists z(PxPy \supset LzxyLzyx)$

i. $\exists z\ x\ y(PxPy \supset LzxyLzyx)$

3. Put these statements into symbols, using the functions you think appropriate where none are offered:

a. Not all Protestants are Methodists. (Construe this *not* as negating the rest of the sentence. Do the same for *b*.)

*b. All Protestants are not Methodists.

c. Some Protestants are not Methodists.

4.*a. Some persons scorn whatever they cannot obtain.

*b. The race is not always to the swift.

c. Nobody believes a liar.

d. Some places are hotter than any frying pan.

*e. All of the boys and girls sang and danced.

*f. There are some elves sitting under that chestnut tree.

g. Something there is that doesn't love a wall.

5. This series is to familiarize you with the effect of the word *only*.

a. If anybody's a Methodist he's a Protestant.

b. All Methodists are Protestants.

c. Only Protestants are Methodists.

d. Only if one is a Protestant is he a Methodist.

e. Only if something is a figure is it a triangle.

f. All triangles are figures.

g. If it's a triangle, it's a figure.

h. All equilateral triangles are equiangular triangles.

j. All equiangular triangles are equilateral triangles.

k. If and only if a triangle is equilateral is it equiangular.

*6. A juggler on tour is well paid if and only if he engages a good agent.

7. a. Perfect love casteth out fear.

b. Write the above statement again, treating *perfect love* as a singular.

8. a. If gods exist, then all is as it should be.

b. If God exists, then all is as it should be.

9. a. If Charlie sees a pretty girl, he whistles.
   b. If Charlie sees a pretty girl, he asks her for a date.
10. a. If something goes wrong, all the staff members complain.
    b. If something goes wrong, some staff member or other sets it right.
11. "Cookie" prepares for each man in the camp any soup that man likes.
12. Some soups are liked by all the men in the camp.
13. There is a cook that prepares any soups in the camp.
14. Every circuit has its own controlling switches, but some switches control all the circuits.
15. In this series add new functions as needed and freely modify the number of arguments in functions already in use:
    a. The man that sells anything will be boycotted.
    b. The man that sells anything to a plant will be boycotted.
    c. The man that sells anything to a plant that employs laborers will be boycotted by every union.
    *d. The man that sells anything to a plant where nonunion laborers are employed will be boycotted by every union.
    *e. The man that sells anything to a plant where nonunion laborers are employed will be thanked by those laborers.
16. Whoever spends all his money discovers his true friends.

| | | | |
|---|---|---|---|
| $Bxy$ | $x$ belongs to $y$ | $Mx$ | $x$ is money |
| $Dxy$ | $x$ discovers $y$ | $Px$ | $x$ is a person |
| $Fxy$ | $x$ is true friend of $y$ | $Sxy$ | $x$ spends $y$ |

17. There's a time for work and a time for play.

| | | | |
|---|---|---|---|
| $Axy$ | $x$ is appropriate for $y$ | $Tx$ | $x$ is a time (occasion) |
| $Px$ | $x$ is play | $Wx$ | $x$ is work |

*18. If any Indians are selling blankets, then no tourist worthy of the name will purchase a blanket from a department store.

| | | | |
|---|---|---|---|
| $Bx$ | $x$ is a blanket | $Tx$ | $x$ is a tourist |
| $Dx$ | $x$ is a department store | $Sxyz$ | $x$ sells $y$ to $z$ |
| $Ix$ | $x$ is an Indian | $Wxy$ | $x$ is worthy of $y$ |
| $n$ | the name of tourist | | |

(Can you make $Sxyz$, despite its three arguments, serve for both the selling and the purchasing?)

19. Turn to the answers (in the back) to question 2, write your own symbolisms for them; then compare these with the original symbolized statements.
20. Prove the two paradigmatic equivalences on page 130.
21. On the basis of the second of these two equivalences, reformulate the symbolism of the first premise at the bottom of page 124.

### EXERCISE IV–8–B

1. Which of these implies the other? Write a proof, using these functions :

   $Cx$ $x$ is a coupling
   $Sx$ $x$ is a size
   $Fxy$ $x$ fits $y$

   a. There are couplings for every size.
   b. Some couplings fit all sizes.

2. One of these implies the other. Write the proof.
   a. Take what librarian you will and what student you will, that librarian finds that student some book or other on some day or other.
   b. For every student there is some day or other when each librarian finds him a book.

3. Some Pitcairn Islanders are descended from mutineers. Mutineers are infallibly persons of enterprise. Therefore, if everyone descended from an enterprising person is himself enterprising, then some Pitcairn Islanders are enterprising.

4. Every chicken snake eats rats. If anything eats anything the latter is killed by the former. Farmer Jones kills every chicken snake he sees. Anything that kills rats is a partner of Farmer Jones. Some chicken snakes are seen by Farmer Jones. So he must kill some of his partners.

5. Anyone who confides in anyone confides in Aunt Helen. Everyone confides in someone or other. Therefore, there is someone in whom everyone confides. (This will be simplified by limiting the universe of discourse to human beings.)

6. Mr. Fenson is a quiet person who delivers fresh milk and eggs to every house on our street and the next. Therefore, every house on our street has milk delivered to it by someone.

7. Here is a problem used by medieval logicians to test the powers of their logical apparatus, not of their intuition: All circles are figures. Hence, anyone who draws circles draws figures.

8. Any history instructor helps any of his students who ask him for help. But none of those helped by a history instructor is his own student. So no student of a history instructor asks him for help.

9. There are some problems that every student attempts. Some students never succeed in what all students attempt. Some teachers help any student that tries something but doesn't succeed. Therefore, some teachers help some students.

## 9. Local Quantification

The whole concept of instantiation is related to the fact that in order to meet the demands of deduction in the quantificational logic, there must be some way of appealing to the propositional. The use of trapezoids represents such a way.[30] The bound-variable procedure gets at the matrices, where the propositional logic is to be found, without removing the quantifiers; nevertheless it takes account of them. Finally, the more rigorous method of instantiation bares the matrices. This method is of wider applicability than the trapezoids and more systematic than the bound-variable procedure, but like each of them it succeeds only by gaining access to the propositional content of the matrices.

In the section following this, the cross-out technique will be adapted to the quantificational logic or, more exactly, the quantificational logic will be fitted with the cross-out technique. Fundamental to such a fitting is some device for turning a quantified matrix into a series of statements amenable to the propositional logic, *i.e.*, something like instantiation must take place, call it by whatever name we will. The device now to be presented serves this purpose (and incidentally a second minor end) and is called local quantification. The name derives from the fact that the transformation herein explained eliminates all quantifiers, yet it preserves *in situ* the distinction between universally and existentially quantified arguments, *i.e.*, the quantification persists only *locally*. This is accomplished by allowing the arguments themselves to preserve their quantifications by writing universally quantified arguments as numerals and existentially quantified ones as letters. An argument that is unquantified, a singular, is written with a letter from the early part of the alphabet as before.

Local quantification is a transformation, resulting as it does in an equivalent statement. It may only be effected, however, on a quantified expression which is itself either asserted or which stands under nothing but dots and wedges. This is to say that a premise such as

$$\exists x(Hx\ y(Jyx \supset Jxy)) \supset xy\exists z(Fxy \supset Gxyz)$$

will, prior to local quantification, be transformed to

$$x(\overline{H}x \vee \exists y(Jyx\overline{J}xy)) \vee xy\exists z(Fxy \supset Gxyz)$$

to eliminate the horseshoe over the quantifiers. In the course of locally quantifying each of these now qualifying expressions, any quantification within a matrix will also be locally quantified. These, too, must therefore qualify.

An expression is locally quantified by these steps:

[30]This is true despite the fact that recourse was had to Venn diagrams in the theoretical exposition of the trapezoid technique.

1. Transform the matrix into an expression having no propositional operators other than dots, wedges, and bars.

This step requires no modification of the first disjunct but rids the second of its horseshoe:

$$x(\overline{H}x \vee \exists y(Jyx\overline{J}xy)) \vee x\ y\exists z(\overline{F}xy \vee Gxyz)$$

2. Rewrite the expression deleting each quantifier. With the deletion of each existential quantifier, replace each variable bound thereby with a letter (the same letter for each occurrence), using a new letter each time such a quantifier is deleted. With the deletion of a universal quantifier, replace each variable bound thereby with a numeral (the same numeral for each occurrence), using a new numeral each time such a quantifier is deleted.

This step yields

$$\overline{H}1 \vee Jx1\overline{J}1x \vee \overline{F}23 \vee G23y$$

Any letter used for a singular, on the other hand, should be used at every recurrence thereof.

Statements such as this locally quantified one are altogether within the reach of the cross-out method, since the only operators are the dot, the wedge, and negations of individual propositions — here functions, which in this locally quantified form will be called *function-propositions* (FPs). The secondary use previously referred to is that two symbolisms which appear different but are really equivalent can generally be seen to be the same when both are locally quantified. Or, if they are much alike but not quite equivalent, the difference between them will frequently be more easily intuited once they are in locally quantified form. In this respect local quantification provides a sort of standard form for quantified expressions in much the same way that DNFs provide a standardized form for those in the propositional logic.

### EXERCISE IV–9

1. Locally quantify the premise and the negated conclusion of the argument 2–a on page 116. Use new arguments, *i.e.*, new letters and numerals, in the local quantification of the negated conclusion, inasmuch as this is part of the same deduction.
2. Locally quantify this argument (negating the conclusion). Be sure to negate the conclusion *before* locally quantifying.

1. $\exists y\ x(HyFx \vee HyGx) \quad \therefore x(Fx \vee Gx \cdot \exists yHy)$

*3. By locally quantifying these pairs of expressions, determine whether they are equivalent or not. Here, because the purpose is to compare the results, you will want to locally quantify the second of each pair *ab initio* instead of using fresh numerals or letters.

a. $\exists x\, Fx \supset p \qquad x(Fx \supset p)$

b. $\exists x\, y\, z(FyGz \supset Hxy) \qquad \exists x\, y(Fy\, \exists zGz \supset Hxy)$

## 10. Instantiation in the Cross-out Technique

An interesting puzzler is whether the locally quantified expression has been instantiated. Since there are no longer any quantifiers, the answer might appear to be an unhesitating yes. If this be true, however, it is somewhat odd that no two FPs, except as they represent recurrences of a bound variable, are instantiated to the same argument. What is more, any establishment of the self-contradiction sought for in the cross-out method must rely, as must any method, on bringing the universal statements to bear on the existential statements, and there is nothing about the locally quantified form *per se* that determines which universals shall be invoked on behalf of which existing elements. In this sense, the instantiation is certainly incomplete. Perhaps the clearest answer is that the existential arguments have been instantiated, each to a different letter as Rule 1 of Section 5 would require, whereas an FP such as *F23* is best said to be semi-instantiated. The numerals serve the function of place-holding as effectively as do the letters but hold in abeyance any specific instantiation, since it yet remains to be seen to what *letter* the numeral *2* or *3* will be applied when the universal statement they embody is invoked.

Invoked it must be, of course, — or, if you will, instantiated completely — in order that the cross-out procedure may go forward, because after local quantification of the negation of the argument

$$1.\ \exists x\, y(Fxy \supset .\ Gx \lor Hy)$$
$$2.\ \exists x\, \overline{H}x \qquad \therefore\ \exists x\, \exists y(Gx \lor \overline{F}xy)$$
$$\sim C\colon\ x\, y(\overline{G}xFxy)$$

has yielded

$$1.\ \overline{F}x1 \lor Gx \lor H1$$
$$2.\ \overline{H}y$$
$$\sim C\colon\ \overline{G}2\ F23$$

the last two lines of which can be conjoined (*cf.* Step 3 of the cross-out method, page 56), giving

$$1.\ \overline{F}x1 \lor Gx \lor H1$$
$$\text{LA} \quad 2, \sim C\colon\ \overline{H}y\, \overline{G}2\ F23$$

— after this is done, the statements that contain the looked-for self-contradiction are at hand, but the self-contradiction is not. The question *What contradicts what?* can be answered only by invoking the universals (numerals) on behalf of certain of the elements declared to exist (the letters). If the instantiation of the numerals were completed by *1* being instantiated to the letter $y$, *2* to $x$, and *3* to the same letter to which *1* is instantiated, namely, $y$, the following could be written:

$$1.\ \overline{F}xy \vee \underline{Gx} \vee \underline{Hy}$$

$$\text{LA} \quad 2, \sim\text{C: } \overline{H}y\,\overline{G}x\,\underline{Fxy}\,\underline{\overline{F}xy}$$

thus bringing to light the self-contradiction which completes the test for validity of the original argument.

At this point it will be profitable to examine the relationship between the cross-out method here emerging and the seven rules governing instantiation and generalization appearing on pages 113f. The feature of this cross-out procedure which may first strike the reader is that although instantiations occur, there is nothing corresponding to generalizations. If it is possible for the cross-out technique to avoid generalizations altogether, would it not be possible to write line-by-line proofs in which no generalizations are resorted to? This is indeed the case. For if any conclusion of an argument be quantified — let us suppose a conclusion to be $\exists x\, y(Fxy\, Gy)$ — then a proof without generalizations could be written in this way:

1. Premise
2. Premise
3. Premise $\quad\quad \therefore \exists x\, y(Fxy\, Gy)$
4. $\text{C} \vee x\, \exists y(\overline{F}xy \vee \overline{G}y) \quad\quad$ GP TQ
5. etc.

If the original argument is valid, the right disjunct of line 4 and the premises of the argument must somewhere contain a self-contradiction. This is to say that the anti-conclusion together with the premises describe a universe which cannot exist. (The reader is invited to recall the solutions worked with bound variables for Section 4.) Now if these describe an impossible universe, it will be because some element of that universe has contradictory statements made about it. Accordingly, the proper instantiations of the premises and this anti-conclusion will pick out such an element and bring to light the two contradictory statements about it. This contradiction, deduced by the aid of the premises from the right disjunct of line 4, will have some form such as $Fx\overline{F}x$ or $Fxy\overline{F}xy$. No generalization will then be needed because the final line, $C$, can be obtained by a TS which omits this self-contradiction.

Such a procedure is unquestionably roundabout when writing line-by-line proofs. That it exists, however, opens the door to proofs without generalizations and so to the technique here contemplated and the more mechanical approach that cross-outs afford. The seven rules can be given a corresponding abbreviation by omitting those concerned solely with generalizations. The remaining ones must persist is some form or another in the cross-out technique. A perusal of page 113 will reveal that Rules 2, 3, and 5 can be immediately dropped from consideration. The others, duly compressed, can now be put more simply:

1 and 4: An EI may not be to a singular or to a representative previously used in that proof for an EI.

6. When one quantifier depends on another, instantiations must be made of both simultaneously or of neither.
7. No instantiation by which any representative becomes its own determinant is licit (even when that determination derives from other instantiations).

The first of these three remaining restrictions has already been cared for. If the local quantification is properly carried out, it is impossible to replace an existentially bound variable with either a singular or a letter that has already been so used. Careless use of the same letter in a second deletion of an existential quantifier when locally quantifying, therefore, is entirely inadmissible because it violates Rule 1 and is thereby fallacious.[31]

Rules 6 and 7 must similarly have counterparts in the cross-out technique. For this reason the procedure shown above (that of replacing the numerals in a cross-out solution with the letters to which they are finally instantiated) *will not be carried out*. For the needed restrictions as to what letters a numeral may licitly be instantiated to will be more easily observed if instead of rewriting the lines as above, *i.e.*,

$$\begin{array}{lr} & 1.\ \overline{F}xy \vee \underline{Gx} \vee \underline{Hy} \\ \text{LA} & 2, {\sim}\text{C}:\ \overline{H}y\,\overline{G}x\,\underline{Fxy}\,\underline{\overline{F}xy} \end{array}$$

we resort to the following way of completing the solution in which the numerals are left standing in their original places in the lines and the final instantiations are indicated by means of a table:

$$\begin{array}{lrcc} & & x & y \\ & 1.\ \overline{F}x1 \vee \underline{Gx} \vee \underline{H1} & 2 & 1 \\ \text{LA} & 2, {\sim}\text{C}:\ \overline{H}y\,\overline{G}2\,\underline{F23}\,\underline{\overline{F}x1} & & 3 \end{array}$$

In the next section these needed controls will be taken up in detail, and it will be seen that the table shown here serves to insure their application. In such a table, *2* will be said to be *assigned* to *x*, and *1* and *3* assigned to *y*; *1* and *3* and *y* will be said to be *co-assigned*. This way of speaking is clearly

[31]The answers to question 3 in the previous exercise reduced two separate expressions to identical local quantifications and in so doing instantiated to the same letter in each operation. But local quantification was there serving only as a device for determining whether one symbolism was equivalent to another. For that secondary purpose, one can afford to set aside the restriction being emphasized here with regard to *proofs* using local quantification.

an alternative to saying that *1* and *3* are to be instantiated to $y$ and *2* to $x$. If the table reveals that the assignments are licit, the completed instantiations appearing in the earlier solution always *can* be written, and knowing this makes the actual writing unnecessary.

The remainder of this section takes up some general controls which have less to do with these four rules than with pseudo-instantiation and the proper way of making repeated instantiations.

If the use of a new letter at each removal of an existential quantifier constitutes an adherence to Rules 1 and 4, what purpose is served by resorting to a new numeral at each deletion of a universal quantifier? The answer to this will be clearer if the following erroneous solution is examined. The inference is that because there are pennies and quarters there exists something that is both.

$$\begin{array}{lll} 1. & \exists x\, Px & \\ 2. & \exists x\, Qx & \therefore \exists x(PxQx) \end{array}$$

Locally quantified:

$$\begin{array}{lll} 1, 2. & Px\ Qy & x \quad y \\ \sim\mathrm{C}: & \bar{P}1 \vee \bar{Q}1 & \end{array}$$

The fallacious 'solution':

$$\begin{array}{llll} \mathrm{LA} & 1, 2. & Px\ \underline{Qy}\ \underline{\bar{Q}1} & x \quad y \\ & \sim\mathrm{C}: & \underline{\bar{P}1} \vee \underline{\bar{Q}1} & 1 \quad 1 \end{array}$$

The error of this pseudo-solution will be manifest if the instantiations indicated by the table are effected. They require that the negation of the conclusion take this form:

$$\sim\mathrm{C}: \bar{P}x \vee \bar{Q}y$$

the error of which corresponds to that in this line-by-line solution:

$$\begin{array}{llll} 1. & \exists x\, Px & & \\ 2. & \exists x\, Qx & \therefore \exists x(PxQx) & \\ 3. & \mathrm{C} \vee x(\bar{P}x \vee \bar{Q}x) & & \mathrm{GP\ TQ} \\ 4. & \mathrm{C} \vee \bar{P}x \vee \bar{Q}y & \text{(a pseudo-instantiation of line 3)} & \end{array}$$

Now the means of preventing such a fallacy in the cross-outs is simply to declare that:

*No numeral shall be assigned to more than one column.*

The question as to why fresh numerals are introduced at each deletion of a universal quantifier can now be answered. If all the numerals were written simply as *1*'s, this would violate no rule, but in view of the rule just laid down it might well cripple the exploitation of universal statements. This can be illustrated by the argument that all cats are sly and all dogs are noisy, Arsenic is a cat and Bowser a dog, so Arsenic is sly and Bowser is noisy.

1. $x(Cx \supset Sx) \cdot x(Dx \supset Nx)$
2. $Ca\ Db$ $\quad\therefore Sa\ Nb$

Were both universals locally quantified without recourse to new numerals:

1. $\overline{C}1 \vee S1$
   $\overline{D}1 \vee N1$

one of the two generalizations would become useless since the numeral *1*, by the above rule, can be assigned to *a* or to *b*, but not to both. By locally quantifying according to the instructions, the solution becomes straightforward:

| | | |
|---|---|---|
| a | 1. $\underline{\overline{C}1} \vee S1$ | |
| b | $\underline{\overline{D}2} \vee N2$ | $a\ \ b$ |
| LA | 2. $Ca\ Db\ S1\ \underline{N2}\ \underline{\overline{N}b}$ | $1\ \ 2$ |
| c | ~C: $\underline{\overline{S}a} \vee \overline{N}b$ | |

One further example will illustrate how the same universal can be invoked on behalf of more than one existing element when required. This argument is that Arsenic and Castor, both being cats, are both sly.

| | | |
|---|---|---|
| | 1. $x(Cx \supset Sx) \cdot x(Dx \supset Nx)$ | |
| | 2. $Ca\ Cc$ | $\therefore Sa\ Sc$ |
| a | 1. $\underline{\overline{C}1} \vee S1$ | $a\ \ c$ |
| | $\overline{D}2 \vee N2$ | $1$ |
| LA | 2. $Ca\ Cc\ S1\ \overline{S}c$ | |
| b | ~C: $\underline{\overline{S}a} \vee \overline{S}c$ | |

At this point the self-contradiction is still lacking because, it being impossible to assign *1* (or instantiate it) to Castor, there is so far no way of invoking the generalization in line 1 on his behalf. The remedy is simple. By rewriting the required generalization with a new numeral, it can become available for assignment to *c*. And with this the self-contradiction is forthcoming.

| | | |
|---|---|---|
| a | 1. $\underline{\overline{C}1} \vee S1$ | |
| | $\overline{D}2 \vee N2$ | $a\ \ c$ |
| LA | 2. $Ca\ Cc\ S1\ \underline{\overline{S}c}\ \underline{S3}$ | $1\ \ 3$ |
| b | ~C: $\underline{\overline{S}a} \vee \overline{S}c$ | |
| c | *R*1. $\underline{\overline{C}3} \vee S3$ | |

An important feature of such rewriting is that nothing less than the *entire original assertion* be rewritten (*cf.* the last line). Were the *S1* brought into the LA from the first line rewritten as *S3*, for example, this would be to repeat in a new form the same fallacy previously made; for it would be treating $x(Cx \supset Sx)$ as if it could rightly be (TH'd and) instantiated to $\overline{C}a \vee Sc$.

A correct rewriting is exactly the same operation as is to be seen in line 4 of this line-by-line proof. A rewrite is simply a second instantiation.

| | | |
|---|---|---|
| 1. | $x(Cx \supset Sx) \cdot x(Dx \supset Nx)$ | |
| 2. | $Ca\ Cc$ | $\therefore Sa\ Sc$ |
| 3. | $Ca \supset Sa$ | 1 UI |
| 4. | $Cc \supset Sc$ | 1 UI |
| | etc. | |

The tabulation of letters below which are placed numerals (which represent the same element as the letter) yields a clear answer to the question about how FPs are struck.

*One FP contradicts another if the function is the same in each except for sign and if argument-place by argument-place, their arguments are co-assigned. Any letter or numeral is obviously co-assigned with itself.*

Despite the fact that the matter has not been explicitly treated, it should be apparent that

*No letter can be assigned to another letter.*

## EXERCISE IV–10

1–2. In the previous exercise (page 136) you wrote the premise and negated conclusion of two arguments in locally quantified form. Complete the solution of those arguments with cross-outs.

*3–4. Write line-by-line proofs paralleling the above cross-out solutions. In making instantiations, be guided by the table of assignments.

5. Use cross-outs to solve the three illustrative problems at the end of Section 5.

*6. For each of the four invalid arguments in question 6, page 103, complete this statement: To effect a contradiction it would be necessary to ______________________ which violates the rules for locally quantifying or for the cross-outs.

7. Show how the error of rewriting with a fresh numeral only part of an assertion can yield a pseudo-proof of the argument about pennies and quarters (page 140).

8. Eleanor remarked of the table of assignments that if *2*, *4*, *7*, and *8* all stood under $z$, there would be five ways (counting $z$ itself) of referring to $z$. Briefly justify her statement.

## 11. The Shifting Variable in Cross-outs

So far, no distinction has appeared between the $Fx1$ that is the local quantification of $\exists x\ y\ Fxy$ and the $Fx1$ yielded by $y\ \exists x\ Fxy$, although such a distinction is plainly indispensable. The safeguards against erroneous instantiations — erroneous assignments, they may here be called — will consist in notations in the table of assignments just as in the development of line-by-line proofs they consisted in the notations of EIs. It is in this table of assignments that the distinction between the two varieties of $Fx1$ is to be made. The $Fx1$ coming from the independent statement, $\exists x\ y\ Fxy$, has no restrictions on it and will therefore go unnotated. The $Fx1$ from $y\ \exists x\ Fxy$, on the other hand, in which the $y$ must not be instantiated to the representative for $x$ — *i.e.*, the *1* must not be assigned to $x$ — calls for a notation of some kind in the table to prevent that assignment. A simple and adequate provision is to enter *1* in the table *above* the $x$; this will serve to indicate that it must not be assigned to $x$, *i.e.*, it must not be entered *below* $x$. According to these conventions, then, the local quantification of $\exists x\ y\ Fxy$ is

$$Fx1 \qquad x$$

whereas that of $y\ \exists x\ Fxy$ is

$$Fx1 \qquad \overset{1}{x}$$

But Rule 7, in whose province these matters lie, concerns itself not only with preventing the instantiation of such a universal to this immediate dependent (which is what the above notation prevents), but also with the illicit 'circuit-closing' of dependences established by further instantiations. The safeguard against this in cross-outs is similar to that resorted to for line-by-line proofs. Let the *1* in the above instance be called a determinant numeral (it so far determines only $x$, of course) and the same designation be used for any numeral so entered above a letter when locally quantifying. Then:

> *Whenever any determinant numeral, N, is assigned to a letter (below the letter, naturally) each determinant numeral standing over that letter is entered above each letter over which N already stands. Any assignment is then illicit if it results in the same numeral appearing both above and below the same letter.*

This procedure prevents precisely those mal-instantiations, or mal-assignments, which Rule 7 prevents. Here is an invalid argument illustrating how the prevention works:

$$1.\ x\ \exists y\ t\ \exists w(Fxywt \lor Gty)$$
$$2.\ y\ \exists t\ \overline{G}ty \qquad \therefore\ \exists t\ x\ \exists y\ \exists w\ Fxywt$$

The first premise is locally quantified thus:

$$1.\ F1yw2 \lor G2y$$

| | |
|---|---|
| | *2* |
| *1* | *1* |
| *y* | *w* |

After the second premise and negated conclusion have been added, other determinants exist:

$$2.\ \overline{G}t3$$

$$\sim\!C\!:\ \overline{F}x564$$

| | | | |
|---|---|---|---|
| | *2* | | |
| *1* | *1* | *3* | *4* |
| *y* | *w* | *t* | *x* |

The assignments are obvious; the only question is whether they are permissible. To place *1* under *x* means that *4* must be entered above the *y* and *w* columns (in each place where *1* already occurs as a determinant). The assignment of *5* to *y* and *6* to *w* causes no ado because neither of these numbers is a determinant. The numbers *2* and *4* are evidently to be co-assigned, but it is not yet clear to what letter, if any. By the time the first disjunct of line 1 has been struck the solution has been carried this far:

| | | | | | | |
|---|---|---|---|---|---|---|
| b | 1. $\underline{F1yw2} \lor G2y$ | | (*4*) | | | |
| | | (*4*) | *2* | | | |
| a | 2. $\overline{G}t3$ | *1* | *1* | *3* | *4* | |
| | | *y* | *w* | *t* | *x* | *2* |
| LA | $\sim\!C\!:\ \overline{F}x564\ \overline{G}t3\ G2y$ | *5* | *6* | | *1* | *4* |

The only question remaining is whether $\overline{G}t3$ and $G2y$ are contradictory. Plainly the (co-assigned) *2* and *4* must now be placed under *t*, which requires writing the determinant *3* over each occurrence of the determinants *2* and *4*. By this time it has become illicit to assign *3* to *y*; no contradiction appears.[32] Note that the provisional entry of *2* and *4* at the right has been cancelled lest these numbers appear in two columns.

[32]This does not prove, of course, that the argument is invalid, but only that the present proof fails to show its validity. Its *in*validity is demonstrated by considering some universe (one of only two elements will serve in the present case) in which the dependent existentials can be selected so as to render the premises true and the conclusion false. The two elements being *a* and *b*, we first make the assertions below.

*Fabaa*
*Gbb*
*Gaa*
*Fbaab*

These will suffice to make the first premise true no matter which elements be chosen for *x* and *t*, provided those elements are taken in the order specified by the quantifiers of that premise. If *x* be *a*, let *y* be *b*; then if for *t*, *a* be selected, the choice of *a* for *w* (resulting in *Fabaa*) renders the first premise true by its first disjunct. If for *t*, *b* is selected, our assertion *Gbb* renders the second disjunct true. Now if *x* be *b*, let *y* be *a*; then if for *t*, *a* is taken,

| | (3) | | | |
|---|---|---|---|---|
| (3) | (4) | | | |
| (4) | 2 | | (3) | |
| 1 | 1 | 3 | 4 | |
| y | w | t | x | ~~2~~ |
| 5 | 6 | 2 | 1 | ~~4~~ |
| | | 4 | | |

The last sentence of the instructions (page 143) for handling determinants, "Any assignment is then illicit if it results in the same numeral appearing both above and below the same letter," clarifies the connection between this procedure and the observance of Rule 7. For if a numeral appears above a letter it is because it determines that letter, and its appearance below that letter corresponds to its instantiation thereto; hence it has become its own determinant.

Only Rule 6 remains to be considered. By examining a violation of it in a line-by-line proof and taking equivalent steps in the cross-out proof, we shall be able to see exactly where caution is required:

*Line-by-line Proof*

| | | |
|---|---|---|
| 1. $x\ \exists y\ Fxy$ | | |
| 2. $\exists x\ \exists y\ z(FxzFyz \supset Gz)$ | $\therefore \exists x\ Gx$ | |
| 3. $x\ Fxy$ | 1 EI (violates | |
| 4. $Fxy$ | 3 UI | Rule 6) |
| 5. $Fwy$ | 3 UI | $w$ |
| 6. $FxyFwy \supset Gy$ | 2 EI UI | $x$ |
| 7. $Gy$ | 4,5,6 c | EI: $y_3x_6w_6$ |
| 8. $\exists x\ Gx$ | 7 EG | |

*Cross-outs*

| | | |
|---|---|---|
| 1. $F1y$ | | 4 |
| 2. $\overline{F}x2 \vee \overline{F}w2 \vee G2$ | | 1 |
| $\sim$C: $\overline{G}3$ | | $y\ x\ w$ |
| R1. $F4y$ | (erroneous) | 2 1 4 |
| | | 3 |

---

the premise is made true by its second disjunct (now $Gaa$), whereas if for $t$, $b$ is taken, let $w$ be $a$ so that $Fbaab$ renders true the first disjunct. In short, these four assertions about a two-element universe insure that the first premise will be true with respect to it. Similarly, we make two more assertions of this universe and the second must be true therein.

$$\overline{G}ba$$
$$\overline{G}ab$$

Finally, the anti-conclusion ($t\ \exists x\ y\ w\ \overline{F}xywt$) will be true — *i.e.*, the conclusion will be false — provided we assert eight more relations in which every case of $y$ and $w$ is covered provided we use $b$ for $x$ when $t$ is $a$ (left column below), and $a$ for $x$ when $t$ is $b$ (right column).

| | |
|---|---|
| $\overline{F}baaa$ | $\overline{F}aaab$ |
| $\overline{F}baba$ | $\overline{F}aabb$ |
| $\overline{F}bbaa$ | $\overline{F}abab$ |
| $\overline{F}bbba$ | $\overline{F}abbb$ |

These fourteen assertions consistently describe a non-empty universe in which the premises of the argument are true and its conclusion false. The argument is therefore invalid, since validity requires that the premises imply the conclusion in *every* non-empty universe.

The pseudo-solution at the right is not completed, but the assignments in the table will easily bring about a self-contradiction. The fallacious solution rests on the mistaken rewriting of line 1. The original line 1 on the right corresponds to line 4 on the left, a line which could be correctly justified by UI and EI from the first premise had not line 3 already been written. *F1y* is similarly licit. But the writing of line 5 on the left is possible only because of the fallacious line 3; and the rewriting of line 1 on the right, which corresponds exactly to this line 5, is equally illicit despite the dutiful entry of *4* as a determinant along with the *1* already standing over *y* in the table. Any regular instantiation of line 1 must instantiate both universally and existentially to avoid a violation of Rule 6. Line 4 could be regularly deduced from line 1; so could *Fwz* and, for that matter, *Fst* and *Fuv* and any number of further instantiations provided a fresh representative were used for each successive EI. The way to rewrite line 1 of the cross-outs lawfully, therefore, is to use a fresh letter along with each fresh numeral. In this way each of these rewritings would be regular:

| | | | | | |
|---|---|---|---|---|---|
| 1. | *F1y* | | | | |
| R1. | *F4z* | *1* | *4* | *5* | *6* |
| R1. | *F5t* | *y* | *z* | *t* | *v* |
| R1. | *F6v* | | | | |

This matter of rewritings, which is indispensable to many proofs and therefore to be mastered, can be further clarified by comparing the solutions to the Adams and Bennett problem on page 122:

| | | | |
|---|---|---|---|
| 1. | $\exists x(Sx \cdot y(Ty \supset Ryx))$ | | |
| 2. | $Ta\ Tb \quad \therefore \exists x(Sx\ Rax\ Rbx)$ | | |
| 3. | $Sx \cdot y(Ty \supset Ryx)$ | 1 EI | |
| 4. | $Ta \supset Rax$ | 3 UI | |
| 5. | $Rax$ | 2,4 c | EI: $x_3$ |
| 6. | $Tb \supset Rbx$ | 3 UI | |
| 7. | $Rbx$ | 2,6 c | |
| 8. | $Sx\ Rax\ Rbx$ | 3,5,7 R | |
| 9. | $\exists x(Sx\ Rax\ Rbx)$ | 8 EG | |

| | | | | | |
|---|---|---|---|---|---|
| a | 1. | $Sx$ | $x$ | $a$ | $b$ |
| | | | $2$ | $1$ | $3$ |
| b | | $\underline{\overline{T}1} \vee R1x$ | | | |
| LA | 2. | $Ta\ Tb\ Sx\ R1x\ \underline{\overline{R}3x}\ \underline{Rb2}$ | | | |
| d | $\sim$C: | $\underline{\overline{S}2} \vee \underline{\overline{R}a2} \vee \overline{R}b2$ | | | |
| c | R1. | $\underline{\overline{T}3} \vee \overline{R}3x$ | | | |

The rewriting of the disjunction from the first premise here uses a new numeral but not a new letter. This is permissible because *1* is not a determinant. Notice that the rewriting corresponds to line 6 in the line-by-line proof, and that any number of rewrites without changing the letter would be licit; they would correspond to repeated instantiations of the general statement in line 3. (In both solutions that generalization is invoked on behalf of both Adams and Bennett with respect to the same student.)

The last precaution required for incorporating the needed rules from the Deductive System into the cross-out technique is this:

> *Nothing less than an entire original assertion is rewritten. If any replaced numeral be a determinant, its dependent letter must also be replaced in the rewriting.*

The argument about the helpful teachers (problem 9, page 134) will prove to be an instructive example of the advantages of the cross-out technique in the quantificational logic.

1. $\exists x(Px\ y(Sy \supset Ayx))$
2. $\exists x(Sx\ y(z(Sz \supset Azy) \supset Fxy))$
3. $\exists x(Tx\ y(Sy\ \exists z(AyzFyz) \supset Hxy))$ $\quad \therefore \exists x\ \exists y(TxSyHxy)$

Once negated, locally quantified, and placed in lines ready for cross-outs, the problem has this appearance:

LA

1. $Px\ Sy\ Tw$
   $\overline{S}1 \vee A1x$
2. $Sz\overline{A}z2 \vee Fy2$ $\qquad$ 2 (above $z$)
3. $\overline{S}3 \vee \overline{A}34 \vee \overline{F}34 \vee Hw3$ $\qquad x \quad y \quad z \quad w$

$\sim$C: $\overline{T}5 \vee \overline{S}6 \vee \overline{H}56$

Notice at this point how many hints the problem offers for its own solution, *i.e.*, toward the correct assignments. On the assumption that the lines are not riddled with redundancies, the polyadic functions occurring only twice, and once negated, are to be trusted to provide indispensable assignments. Consider the *F* function. If the sole two occurrences thereof are to contradict each other, *3* must be assigned to *y* and *2* and *4* must be co-assigned. The *H* function requires that *5* be assigned to *w* and that *6* be co-assigned with *3*, *i.e.*, under *y*. The *A* function suggests that *4* (and with it the co-assigned *2*) be placed under *x*. (This licit assignment of *2* occasions no entries above the letters because no determinant stands over *x*.) The *1* of the *A* function in the first premise must be assigned either to *z* or else co-assigned with the *3* already under *y* if that function is to cross either of its negated forms in premises 2 and 3. This decision can be deferred until the cross-outs already possible have been effected, but even then it must be made on a trial and error basis. On the proper assignments, $Sz\overline{A}z2$ can eventually be added to the LA. A rewriting of the disjunction in premise 1, using a new numeral which is then assigned to *z*, achieves the required contradiction.

Many times such exploitation of indications within a problem affords all the assignments. But when assignments must be guessed at, or discerned apart from such indications, or when rewritings such as the above are required, then of course the method ceases to be effective in the technical sense. Indeed, no method is here effective. It is to the credit of cross-outs that the method reduces to a minimum the demands on intuition and creativity.

### Rules for Cross-outs in the Quantificational Logic

One FP contradicts another if the function is the same in each except for sign and if in each argument-place their arguments are co-assigned. P. 142

In making assignments:

a. No numeral shall be assigned to more than one column, nor any letter assigned to a letter. Pp. 140, 142

b. Whenever any determinant numeral, *N*, is assigned, each determinant numeral standing over that column is entered above each column over which *N* already stands. Any assignment is then illicit if it results in the same numeral appearing both above and below the same letter. P. 143

In rewriting local quantifications, any numeral (or letter) may be replaced at each occurrence by a new one, provided

Nothing less than an entire assertion is rewritten. If any replaced numeral be a determinant, its dependent letter must also be replaced in the rewriting. P. 147

## EXERCISE IV–11–A

1. Use cross-outs to prove the expressions in question 10, page 122.
2. Complete the solution of the illustrative problem at the end of the present section.
3. In view of what was explained on page 143, it is plain that the local quantification arrived at in Section 9 is incomplete. Set down that local quantification in completed form.

*4. As an aid to understanding the device governing determinant numerals, attempt to solve the invalid as well as the valid arguments in question 1, page 116.

5. By the device of a universe of a small number of elements (*cf.* the note on page 144) show that these arguments are invalid:
   a. 1. $x\,\exists y\; Hxy$ $\quad\therefore \exists x\; Hxx$
   b. 1. $x\,\exists y\; Mxy$ $\quad\therefore \exists y\; x\; Mxy$
   c. 1. $x\,\exists y(Fx \equiv Gy)$ $\quad\therefore \exists y\; x(Fx \equiv Gy)$
   *d. The argument on page 145.
   e. 1. $x\,\exists y\; z\,\exists w\; Fxyzw$ $\quad\therefore x\,\exists y\; Fxyyx$

6. Jill believed there would be nothing amiss about making a new table of (different) assignments for the second branch of a solution having a branched LA. Jan feared that unless the same table of assignments were used throughout a proof, the proof would be fallacious. She succeeded in proving her point using the penny-and-quarter argument (page 140). Write out the pseudo-proof which she showed Jill.

## EXERCISE IV–11–B

1. Using cross-outs, prove each of the following:
   a. 1. $x\,\exists y\; Fxy$ $\quad\therefore x\,\exists y\,\exists z(FxyFzy)$
   b. 1. $x\,\exists y\; w(Fxyw \vee Gxw)$
      2. $x\,\exists y\; \overline{G}yx$ $\quad\therefore \exists y\,\exists w\,\exists x\; Fxyw$
   c. 1. $x\,\exists y\; Fxy$ $\quad\therefore x\,\exists y\,\exists z(FxyFyz)$

2. From Exercises IV–7–B and IV–8–B, pick out the five problems which gave you the most difficulty. Solve them by cross-outs.

3. For the following problems use either cross-outs or line-by-line proofs. Do not mix the methods in any one problem, of course.
   a. 1. $x(KxBx \supset \overline{F}cx)$
      2. $Ba$ $\quad\therefore Ka \supset \overline{F}ca$

   b. 1. $x\,\exists y(FxGy \vee \overline{J}xHy)$
      2. $\exists x(GxHx)$ $\quad\therefore \exists y\; x(FxGy \vee \overline{J}xHy)$

   c. 1. $x\,\exists y(Txy\; Py)$
      2. $\exists x(Px\; y(Ryx \supset \overline{T}xy))$ $\quad\therefore \exists x\,\exists y\; \overline{R}xy$

4. Use cross-outs to show the equivalence (by deducing each from the other) of Tim's and Tom's writings. They were:

$$x\,\exists y(Cx \supset Py \cdot Px \supset Cy) \qquad \exists x\; Cx \supset \exists y\; Py \cdot \exists x\; Px \supset \exists y\; Cy$$

## SUMMARY

The symbolism of the quantificational logic can be evolved from that of the propositional by treating what were unconnected propositions as predicates of a common subject, which then becomes the argument of each predicate or function. The argument is quantified universally to mean that the statement is true of every element of the universe, existentially to mean that it is true of some.

The Aristotelian Square of Opposition, because of the existential import of its universal propositions, entails implications which, except for contradiction, are lost in the interpretation given modern symbolism. From the modern square can be deduced the transformations TQ.

Trapezoids prove useful for testing validity provided the functions involved are monadic and singly quantified. The explanation of why the method is effective is grounded in the Venn diagrams, thus illustrating a community of interpretation in the case of two symbolisms ordinarily thought of as distinct.

The bound-variable procedure (so called because it leaves the quantifiers in their places) provides for testing many arguments by leaning heavily on intuition rather than manipulation. This reliance on intuition is supplanted by the rule-governed use of instantiation and generalization. By expanding Rule II of the Deductive System to include these operations, the quantificational logic is made to rest on the propositional. For once quantifiers are stripped away by instantiation (later to be restored as needed by generalization), the contents of matrices become simply so many true-or-false statements amenable to the treatment accorded propositional variables. The last three rules governing these new inferences are concerned with dependent variables. The distinction between existentially quantified variables that are determined and those that are not, together with the exploitation of the flexibility of Rule II, makes the scope of permissible inferences much wider than in systems commonly in use.

The advent of quantifiers elicits three new transformations by which Rule I is expanded. The distribution of quantifiers (TQD) and the legitimate changes in their scope (TQS) afford insight not only into the nature of the quantifiers but also into the equivalence of variously rendered symbolizings.

The process of correctly symbolizing propositions expressed in colloquial language can be formulated only to a limited extent; it is largely a matter of capturing meanings rather than the linguistic symbols conveying them, and of facility with the logical symbols. Five informal guide lines are offered as helps.

Local quantification is at one time a standard form into which to cast diversely written quantified expressions and also a way of making the quantificational logic accessible to the cross-out method. Existential instantiations are effected in the process of locally quantifying; universal instantiations are effected partly thereby and then completed, so to speak, by the table of assignments. Because solutions by cross-outs entail no generalizations, only the rules governing instantiations must be incorporated in the cross-out method. By following regular procedures of writing and rewriting local quantifications and of making assignments, the restrictions of the Deductive System are automatically observed. Although no method is effective for all polyadic functions, cross-outs appear to be as great a boon to the quantificational logic as to the propositional — partly because the 'semi-instantiation' of universals to numerals provides a convenient distinction between what can and cannot be invoked on behalf of existential statements and what arguments can and cannot oppose others in contradictory FPs, and partly because the assignments necessary for a solution are often indicated by the FPs themselves.

# V

# Further Aspects of Quantificational Logic

## 1. Identities

One bit of symbolism goes a long way in enlarging the quantificational logic: $a = b$ (read *a is identical to b*) is the symbolism for affirming that $a$ and $b$ are the same thing. The components of the identity are not classes or propositions, but *elements* (or more strictly, an element) in the universe of discourse. The statement is not about a group of things, despite any quantification under which it may be found; $a$ stands always for *it* or *that thing*, in the singular, and of course $b$ stands for the same thing differently designated. Even the contradictory statement that $a$ is different from, or not identical to, $b$ ($a \neq b$) has individual elements for its components.

The first expansion afforded by this addition is that the language can now refer to singulars in this way. *Joe Smith is really Joseph Bernard Smith* might be symbolized $j = s$; *that girl over there is Jane* can be expressed as $g = j$. The essential feature of identity — the key to the sameness of one thing designated in two different ways — was defined by Leibniz and probably by others before him. It is that a thing is itself because nothing else shares every last one of its properties; *i.e.*, if 'two' things were alike in *every* respect, they would not be two, but one only. Two grains of sand, or two hydrogen atoms, or what-have-you, between which we might hastily suppose we have no way of distinguishing, are still two precisely because this one is over here and that one is there so that location is at least one property they do not have in common. Everything considered, this is a pretty safe doctrine! [1]

[1] Among the concerns of psychiatrists is a phenomenon called paralogia, a manner of thinking unrelated to ordinary 'logical' thought. This is not to say, however, that the paralogic individual does not have his own way of being logical. It can often be seen that

This community of properties of 'two' identical things is made to order for the language of quantificational logic, for it means that once $a$ is identified with $b$, any functions of either can be deduced to be functions of the other, whether the function be affirmatively or negatively expressed. By the same token, once it is known that there is some function which $a$ and $b$ do not share — however many other properties they may have in common — it can be deduced that $a$ is not $b$: $a \neq b$. This can be incorporated into the Deductive System under Inferences:

f. $x = y$ implies that any function of either $x$ or $y$ is a function of the other.[2]

f′. $Fx$ and $\overline{F}y$, both being asserted, imply $x \neq y$.

This symbol adds a second 'improvement' to the quantificational logic: the ability to say *how many* of something exist. The symbol for existence, $\exists x$, means only that the statement it precedes is true of at least one element in the universe. It might in fact be true of only one element, or of twenty or twenty million; up to now no distinction between these alternatives has been possible. To say that there are (one or more) apples on the table, the symbolism provides for saying that there exists an $x$ — perhaps dozens of them — such that $x$ is an apple and $x$ is on the table: $\exists x(AxOxt)$. With the help of identity, the same language can now state that there exists one and only one apple on the table, *i.e.*, there exists an $x$ such that $x$ is an apple and on the table and *anything in the universe that is an apple and on the table is identical to this x referred to:*

$$\exists x(AxOxt\ y(AyOyt \supset y = x))$$

Although it is impractical to undertake with this richer symbolism to say that there are twenty apples on the table, such a statement is theoretically within its scope. The following statement:

$$\exists x\ \exists y(AxOxtAyOyt\ z(AzOzt \supset . z = x \vee z = y))^3$$

---

he is *identifying* things by the fact that they have some common property; thus one thing is not only like the other but *is* the other. For most of us the universe is made up of elements thought of as *things*, so that *same* means the same thing, and if we say *same* of a property, as in "That's the same temper his Daddy used to display," we mean only to point out a similarity between elements. The paralogic's universe is apparently comprised instead of properties, or predicates, so that he may answer the question, "Who is the President?" with "White House," because living in the White House is a property of the individual who is the President. By the same token, the paralogic's speech is commonly rife with metaphors, the difference between his metaphors and ours being that very often his are not figures of speech but identities.

[2]Here the word *function* embraces identities and non-identities themselves. That is, given $x = y$, the inference under $f$ provides that if $x = z$, then $y = z$, or if $x \neq w$ then $y \neq w$. In $f'$, if $x = z$ and $y \neq z$, then the function of being identical to $z$ is one which $x$ and $y$ do not share; that is, it can be inferred that $x \neq y$.

[3]The rank of the new operators ('=' and '≠') with respect to the dot, wedge, etc., is trouble-free; even the unwritten conjunction must always rank *above* them. Because the components of identities and non-identities are always *elements*, no operator we have studied can rank lower.

does *not* say that there are two apples on the table, but that there are no more than two. There might still be only one if $x$ and $y$ are, after all, the same element. To say that there are two requires saying not only that there is no third but also eliminating the possibility of there being but one:

$$\exists x\, \exists y(AxOxtAyOyt\; x \neq y\; z(AzOzt \supset .\; z = x \vee z = y))$$

It might appear at this point that number theory has its logical foundations here. This is not the case. The logical theory on which number theory rests is more elaborate than anything found in this text. Nevertheless, being able to write that there are exactly so many of something is a great advance. In ordinary language, this new ability to state that there exists only one thing of a particular description opens the door to a greater degree of detail and faithfulness in translating certain linguistic expressions into symbolism.

To be sure, singulars have been a part of the symbolism since its inception, so that there has long been a way of writing such an expression as *The man who shot Kennedy was called Oswald* by treating the man who shot Kennedy as a singular element and *called Oswald* as a function and so writing $Om$. Using identity it is possible to write $m = o$, but this really says nothing more than does $Om$, as it fails to pick up the peculiarities of the linguistic construction *the man who shot Kennedy*. Russell, one of the pioneers in investigating this point, gave to such phrases the name *definite descriptions*. The definite article, so used, imparts a peculiar meaning: the phrase would be inappropriate if there were two men who shot Kennedy or if there were no man at all who did so. Accordingly, such definite descriptive phrases connote the existence of some element to which the phrase is suited and also the uniqueness of that element. The use of identity provides the expression of these meanings:

$$\exists x(MxSxk\; y(MySyk \supset y = x))$$

This is a *statement* to the effect that there is a man — just one — who shot Kennedy, whereas the descriptive phrase is grammatically only a nominative clause; the point is that the clause is appropriate only because the statement is judged true. Whatever else is expressly predicated of the subject referred to by the nominative clause, *e.g.*, that he was called Oswald, is easily integrated with the symbolized statement for the definite description:

$$\exists x(MxSxk\; y(MySyk \supset y = x)Ox)^4$$

This third benefit is widely adaptable. To refer to the ablest man on the team, a statement is made to the effect that there is one team member more able than all other members of the team:

$$\exists x(Tx\; y(Ty\; y \neq x \supset Axy))$$

Two things are to be noted here. This, again, is a statement implied by the nominative phrase; because the latter is a definite description, a statement is

[4]In the place of $Ox$ it would be possible to write another identity, of course: $x = o$, $o$ being a singular element.

required. As before, any further statement about the individual is to be incorporated into this statement by conjunction. The second thing is that the non-identity in the antecedent is essential. Were it omitted, the meaning would be that this individual is abler than any member at all, including himself.

By way of summarizing, let the definite description be contrasted with the indefinite descriptive phrase. *The man who shot Kennedy* and *the man who wants to get ahead* are both nominative clauses. The first has existential import and is therefore symbolized by a matrix prefixed by $\exists x$. The second lacks such import (not in point of empirical fact, to be sure, but in point of meaning) and consequently merits no existential quantifier but rather a universal one which allows for the possibility of the class being empty. Finally, the *the* in the one phrase has an entirely different meaning from the same word in the other. An initial *the* which introduces neither a singular nor a definite description generally has the sense of *every*.[5]

In the cross-out method, statements of identity and of non-identity are to be regarded simply as a new variety of FP. Obviously $x = y$ contradicts $x \neq y$. The only irregularity — easily cared for — comes from the fact that once identities are introduced, a contradiction may consist in three FPs taken together, *e.g.*, $x = y$, $Fx$ and $\overline{F}y$, or $x = y$, $x = 4$ and $y \neq 4$. If such a triad of FPs appears in the LA, they constitute a self-contradiction just as surely as do $Fx$ and $\overline{F}x$. By the same token, any two of them asserted in the LA are sufficient for striking a wedged disjunct in which the third appears. In the following examples, the underscored disjuncts are struck by the FPs in the LA:

(1) LA $Fx \ \ x = y$
$Gx \vee \underline{\overline{F}y}$

(2) LA $Fx \ \ \overline{F}y$
$Gz \vee \underline{x = y}$

(3) LA $x = y \ \ x = 3$
$Gz \vee \underline{y \neq 3}$

But unfortunately it is not enough to admit such triads as inconsistent. Here is an inconsistent tetrad: $x = y \ \ y = z \ \ z = w \ \ x \neq w$. Five statements can provide such a self-contradiction: $Fxyz \ \ \overline{F}stw \ \ x = s \ \ y = t \ \ z = w$. For this reason it is more convenient to use some convention for *reducing such contradictions* to that between two FPs by modifying one of the three, or working two modifications on an inconsistent tetrad, or three on a group of five statements, and so forth. Taking $Fxy \ \ \overline{F}wz \ \ x = w \ \ y = z$ as an example, the first FP can be modified by virtue of the third, thus:

$$F\underset{w}{\not{x}}y \quad \overline{F}wz \quad x = w \quad y = z$$

and by virtue of the last FP a second modification can be made in either the first or the second:

[5]An inverted iota can be used instead of the existential quantifier to symbolize the definite description: ℩$x\ Fx$ is read *the (unique) x such that Fx*. The advantage at this level of the more extended symbolism is that it is comprised of manipulable FPs.

$F\not{x}\not{y}\ \overline{F}wz\ x = w\ y = z$ or $F\not{x}y\ \overline{F}w\not{z}\ x = w\ y = z$
$wz$ $w \quad y$

With this done, the self-contradiction has assumed the usual form, namely, two FPs have co-assigned arguments and differ in sign.

By this convention, the proper ways of striking the disjuncts shown above would be these:

(1′) LA $Fx\ x = y$ (2′) LA $Fx\ \overline{F}y$ (3′) LA $x = y\ x = 3$
$Gx \lor \overline{F}\not{y}x$ $Gz \lor \not{x = y}Fy$ $Gz \lor y \neq \not{3}x$

Observe that in 2′ an identity has itself been modified to become a conventional function. The change is based on the asserted $Fx$. The resultant $Fy$ is struck by the asserted $\overline{F}y$ in the usual way. For a wedged disjunct which is part of a self-contradictory tetrad to be thus modified and struck, the other three FPs of the tetrad would have to be asserted. If the self-contradictory group is comprised of $n$ FPs, $n - 1$ of them must appear in the LA in order for a wedged disjunct containing the other member of the group to be crossed.

Inasmuch as any element of the universe represented by a letter (or by a numeral) is as well designated by each numeral co-assigned therewith, the following modifications are equally legitimate, supposing the table at the right to be a part of the solution:

LA $F3\ x = 4$ LA $Fy\ \overline{F}3$ LA $6 = 3\ 2 = 5$
$Gz \lor \overline{F}\not{2}x$ $Gz \lor \not{1 = 4}F1$ $Gz \lor 4 \neq \not{x}6$

| $y$ | $x$ | |
|---|---|---|
| 2 | 3 | 5 |
| 4 | 1 | 6 |

It is well to note that the above modifications would be no less licit even if the numeral $4$ stood over $x$ in the table as a determinant. Such a notation means only that in the statement whose writing occasioned it (perhaps $z\ \exists w\ Fzw$, for example) nothing warrants assigning $4$ to $x$; it is entirely possible, of course, that other statements about the universe should warrant that very assignment. An example: $x\ \exists y(FxyGy)$ does not justify the assumption that anything is $F$ to itself. If conjoined to this, however, is the statement that there exists but one $G$, $\exists x(Gx\ y(Gy \supset y = x))$, then that particular element which is $G$ must be $F$ to itself.

These modifications yielding a conventional contradiction between two FPs correspond, as far as theory goes, to the schemata $f$ and $f'$ added previously to the Deductive System. They can be incorporated into the cross-out technique by adding this rule:

> *If an identity is asserted, any FP having an argument co-assigned with one of its components can have that argument replaced by the other component. An identity occurring in a wedged disjunct can be replaced by any asserted FP having an argument co-assigned with one of its components provided that argument is replaced by the other component.*

When one (or both) of the components of an identity is a numeral, it should be assigned so as *not* to be co-assigned with the other component. A little reflection will reveal the reason: if the two components of an identity are co-assigned, the identity is rendered useless because the table already avers what the identity is capable of asserting. For example, given these lines of a solution and this table:

| LA | $\overline{G}x\ \overline{G}w\ Fwx$ | $x$ | $w$ |
|---|---|---|---|
| | $\underline{G2} \vee x = 2$ | $2$ | |

notice that even after $G2$ is struck and the identity transferred to the LA, nothing has been won which is not already provided by the table itself. By assigning $2$ to $w$, the identity is equally well won for the LA, but is not rendered vacuous.

Here is a simple proof respecting identities. Its premise, construing $Vxy$ to mean that $x$ is a function characterizing $y$, is the symbolizing of the definition given earlier of the notion of identity.

| | | |
|---|---|---|
| 1. | $xy(z(Vzx \equiv Vzy) \equiv x = y)$ | Premise |
| 2. | $z(Vzx \equiv Vzx) \equiv x = x$ | 1 UI |
| 3. | $Vzx \equiv Vzx$ | GP |
| 4. | $zx(Vzx \equiv Vzx)$ | 3 UG |
| 5. | $z(Vzx \equiv Vzx)$ | 4 UI |
| 6. | $x = x$ | 5,2 I |
| 7. | $x(x = x)$ | 6 UG |

From the fact thus established that everything is identical to itself, it follows that $x \neq x$ is self-contradictory and can be treated the same as $p\overline{p}$. Moreover, this regulation for cross-outs is justified:

*A non-identity is rendered false provided its components are (licitly) co-assigned.*

It will sometimes be advantageous to exploit this means of falsifying a non-identity. On other occasions, as the exercise will also show, more is to be gained by keeping the components of non-identities in separate columns, the better to cross identities.

## EXERCISE V–1–A

This exercise introduces material not to be found in the text proper.

1. Symbolize each of these phrases or sentences:

a. The acting mayor
b. The acting mayor is a capable man.
c. The acting mayor is the most capable man on the council.
d. There are at most two authorities worth heeding.
e. There are two parts to Pakistan.
f. Three students got top honors.
*g. There are two or three persons in the room. (Use $Rx$ for *x is a person in the room.*)
*h. The most ambitious student needs adequate rest.
j. The most avid student will be rewarded.
k. Tommy eats more than anyone else in the family.
*l. The tallest debater is not Harry.
*m. Nobody but Tom can do that.

2. Prove these arguments, using cross-outs:

a. 1. $Ga$
2. $\overline{G}c \quad \therefore c \neq a$

b. 1. $x(Px \equiv Qx) \quad \therefore x\, y(Px\bar{Q}y \supset x \neq y)$

c. 1. $\exists x(HxJx)$
2. $x(Jx \supset \exists y\, Fxy)$
3. $\exists x(Hx\; y\, \overline{F}xy) \quad \therefore \exists x\, \exists y(HxHy\; x \neq y)$

d. 1. $x(\overline{A}x \supset BxCx)$
2. $\exists x(Bx\overline{C}x)$
3. $\exists x(\overline{B}xCx)$
4. $\exists x(\overline{B}x\overline{C}x) \quad \therefore \exists x\, \exists y\, \exists z(AxAyAz\; x \neq y\; y \neq z\; x \neq z)$

3. a. Is $Fx\; Fy\;\; x \neq y$ a self-contradictory triad?
*b. Complete this statement: Of any self-contradictory triad such as this section treats of, at least one FP must be ____________________.

4. Show that $x\; \exists y\;\; x = y$ is logically true, using cross-outs.

5. Is assigning its arguments to different columns enough to make an identity false? Test your surmise by attempting to prove the statement $\exists y\;\; x\;\; x \neq y$. Does your answer endorse the proof or prevent it? Is the statement a tautology, a self-contradiction, or a contingency?

6. a. As to the statement $\exists y\;\; x\;\; x = y$,
1. Can you specify a universe in which this statement is true?
2. A universe in which it is false?
3. Is the statement a tautology? A self-contradiction? A contingency?
4. Can a null-premise proof of this statement be valid?

*b. Undertake to prove the above statement by cross-outs in order to see how this question should be answered: Can a determination exist in a non-identity?

7. Here are two arguments with proofs purporting to be valid. After intuiting what the argument alleges and carefully examining the proof, state the proof to be valid or else cite the rule it violates.

a. 1. $\exists y\ x\ x = y \quad \therefore y\ x\ x = y$

LA 1. $\underline{1 = y}\ 2 = y\ \underline{x \neq y}$

b $\sim$C: $x \neq \cancel{z}y \qquad y \quad x \quad z$

a R1. $2 = y \qquad\qquad 1 \quad 2$

b. 1. $\exists x\ Gx$

2. $\exists x\ \overline{G}x \quad \therefore \exists y\ x\ x \neq y$

LA 1. $Gx\ \overline{G}y\ \underline{Gz}\ \underline{\overline{G}z} \qquad 2$

a 2. $\overline{G}y \qquad 1$

b $\sim$C: $\cancel{z =}\ 1Gz \qquad x \quad y \quad z$

c R $\sim$C: $\cancel{z =}\ 2\overline{G}z \qquad 1 \quad 2$

### EXERCISE V–1–B

Prove the following arguments:

1. Everyone speaks well of some talented person or other who is widely known. Only one widely known person is talented. So someone speaks well of himself.

2. 1. $a \neq b \quad \therefore x\ \exists y\ x \neq y$

3. 1. $x\ y\ x = y \quad \therefore \exists y\ x\ x = y$

4. Charles has a pet that is noisy. He has a pet snake. No snakes are noisy. So it is false that he has just one pet.

5. 1. $\exists x\ Fxx$

   2. $x\ y(Gxyx \supset x = y)$

   3. $x\ y(Gxxx \supset Fxy)$

   4. $x\ y(Fxy \supset \exists w\ Gxwy) \quad \therefore \exists x\ y\ Fxy$

*6. Everyone talks to everyone about someone or other who is either patient or noteworthy. One is noteworthy if and only if he is patient. There's only one patient person. Therefore, everyone is talked to about someone or other by everyone.

*7. (In attacking this one, remember that you need to change a letter in a rewriting only if you are changing a numeral determining it.)

   1. $x\ \exists y(Bx \supset Fyx\ w(Fwx \supset w = y))$

   2. $\exists x\ \exists y\ \exists z(Fxz\ Fyz\ x \neq y\ y \neq z\ z \neq x) \quad \therefore \exists x\ \overline{B}x$

8. Everything is $F$ to something or other. Nothing is $F$ to itself. If anything is $F$ to anything, the latter is not $F$ to the former. So there must exist at least three things.

*9. If everyone at the party talks gaily with everyone, the hostess will feel at ease. The hostess doesn't feel at ease because everyone is talking gaily with the girl in the red wig. Some people at the party simply don't talk gaily with anyone that doesn't feel at ease. Therefore, the girl in the red wig is not the hostess and someone at the party is not talking gaily with someone. (A singular is sufficient for *the girl in the red wig.*)

## 2. Properties of Predicates

The argument that because all the prophets were persecuted, Elisha must have been persecuted, too, is a sound one. But it would be impossible to demonstrate it without the explicit statement of a premise which is implied only, namely, that Elisha was a prophet. Such a premise is not stated because it is meant to be understood or is already known. In any case, it is taken for granted. Such an unstated premise is called *enthymematic;* the argument is termed an *enthymeme.*

The enthymematic premise may be a statement of fact, as in the above example. It may also, however, be a statement about language. That Elmer is taller than George because he is taller than Frank and Frank is taller than George lacks a premise of this sort. This argument cannot be demonstrated without expressing the premise that what is taller than something taller than a third thing is itself taller than that third thing. *Cincinnati is north of Mobile because it's north of Nashville* has two unstated premises: one is a matter of fact (that Nashville is north of Mobile), the other a matter of language (that what is north of what is north of something is north thereof).

The factual enthymematic premise introduces nothing of special interest. The linguistic premise is frequently a general statement *about a function.* The two examples just given, *being taller than* and *being north of,* are *transitive* functions. Many predicates have this property, the generalized statement of which is this:

$$x\ y\ z(FxyFyz \supset Fxz) \qquad \text{Transitivity}$$

This formulation defines transitivity: a function is transitive if and only if this statement is true with respect to it. The statement is false of many functions, of course; *ancestor of* is transitive but *mother of* is intransitive. There are other functions which are not necessarily the one or the other (*friend of,* for example). One of the most important transitive functions is class-inclusion, for this transitivity is fundamental to class logic.

The way to introduce an unstated premise that a particular function is transitive is to assert transitivity of that function. The conclusion that Elmer is taller than George is reached in this fashion:

| | | |
|---|---|---|
| 1. | $Tef$ | |
| 2. | $Tfg \quad \therefore Teg$ | |
| 3. | $x\ y\ z(TxyTyz \supset Txz)$ | unstated premise |
| 4. | $TefTfg \supset Teg$ | 3 UI |
| 5. | $Teg$ | 1,2,4 c |

That Cincinnati is north of Mobile rests on these premises:

1. $Ncn$
2. $Nnm$ — unstated premise
3. $x\ y\ z(NxyNyz \supset Nxz)$ — unstated premise

To say that a function (*e.g.*, *mother of*) is *intransitive* is to say not merely that there exist mothers, the mothers of whose mothers are not their mothers; for such a statement, $\exists x\ \exists y\ \exists z(MxyMyz\overline{M}xz)$, merely denies that the relation is transitive. To state that the relation is intransitive requires an expression just as universal as that of transitivity:

$$x\ y\ z(MxyMyz \supset \overline{M}xz) \qquad \text{Intransitivity}$$

This alleges that in no case is *being mother of* a transitive relation. The remark that *friend of* is neither transitive nor intransitive simply means that of this relation one can correctly affirm neither the one statement nor the other. Neither generalization is true of it. The exercise will open the question of whether it is possible for a function to be both transitive and intransitive. If such a function exists, it must be one of which both statements can be safely made. Can you anticipate what sort of function this must be?

Another property assignable to some predicates is *symmetry*. *Sibling of* exemplifies this, which is generalized in this way:

$$x\ y(Fxy \supset Fyx) \qquad \text{Symmetry}$$

*Mother of* is *asymmetrical*, *i.e.*, of this relation we can correctly affirm:

$$x\ y(Mxy \supset \overline{M}yx) \qquad \text{Asymmetry}$$

With respect to the relation *friend of* neither generalization is true, so that this relation is neither symmetrical nor asymmetrical.

Still another property of predicates is reflexivity:

$$x\ y(Fxy \supset Fxx) \qquad \text{Reflexivity}$$

*Co-assigned with*, in the cross-out method, has this property. *Being in the same room with* is another example. *Mother of* is irreflexive. Many functions (*friend of*, *defender of*, etc.) are neither.

A relation that is transitive, symmetrical, and reflexive is said to place its arguments in a *relation of equivalence*. For example, if in the domain of plane figures $Cxy$ means that $x$ is congruent with $y$, then with respect to their shape all congruent figures are equivalent. Similarly, within the domain of integers, 1, 4, 7, 10, . . . (*i.e.*, $3n + 1$) are equivalent with respect to their remainder after they are divided by 3 (the integers 2, 5, 8, etc., being similarly equivalent to each other; 0, 3, 6, etc., constituting the third set of equivalent integers) since if $Rxy$ means that the remainder of $x$ when divided by 3 is equal to that of $y$ when so divided, this relation qualifies as an equivalence relation.

## EXERCISE V–2

*1. Write the generalized formula for irreflexivity.

2. Are these formulas logical truths or contingencies?

3. Classify each of the following with respect to each of the three properties in the way the first is done. If it seems to you that a relation has both a property and its contrary, so mark it.

| | Tr. | Int. | Neither | Sym. | As. | Neither | Ref. | Ir. | Neither |
|---|---|---|---|---|---|---|---|---|---|
| a. blood relative of | | | X | X | | | X[6] | | |
| b. cousin of | | | | | | | | | |
| c. grandson to | | | | | | | | | |
| d. square root of | | | | | | | | | |
| e. heavier than | | | | | | | | | |
| f. younger than | | | | | | | | | |
| g. spouse of | | | | | | | | | |
| h. wife of | | | | | | | | | |
| j. in the presence of | | | | | | | | | |
| k. benefactor to | | | | | | | | | |
| l. multiple of | | | | | | | | | |
| m. sympathetic to | | | | | | | | | |

Which of the above relations is an equivalence relation?

4. Starting with this statement about a relation: $x\ y\ z(Fxy \supset \overline{F}yz)$, see if you can deduce transitivity of the relation. Can you deduce its intransitivity? Does this affect your answer to 3–h?

5. In solving these problems, resort where necessary to unstated but understood premises, whether factual or linguistic.

   *a. Alfred has better grades than Tony, who stands in nineteenth place. So you know he has better grades than whoever is in fortieth place. (Use $Nx$ for $x$ is in nineteenth place, and so on.)

   b. He's not in the same room with himself; so he's not in the same room with anybody.

   c. Nobody is Sue's spouse, so nobody's spouse is she.

   d. Anything called a hill in Utah is higher than anything in the Adirondacks. Anything in the Adirondacks is higher than the 'Florida Alps.' Therefore anything called a hill in Utah is higher than the 'Florida Alps.' (Be prepared to add an indispensable *factual* premise.)

[6]Whether some functions are reflexive or not is debatable. Is a man his own relative? His own compatriot?

e. The predicate $K$ is transitive. It is also irreflexive. Hence, it is asymmetrical.

f. Said Tate to Tote, "The same mother bore us, the same father begot us, but we are not siblings, nor am I thee." Therefore, Tate was a liar. (You will need, of course, a symbolized definition of *sibling of.* To write this correctly, keep in mind Tate's last four words. Would he necessarily be lying if he omitted these words?)

*g. Our class has officers of unlike sex. There are also officers of the same sex. So there must be at least three officers.

## 3. The Nature of Axioms

Up to this point the concern of every explanation has been one or another method. One section, even so, took up a proof that the cross-out method is efficacious in the propositional logic for distinguishing between valid and invalid arguments. Even if logicians' sole concern were validity, they would very naturally move up to another level of systematization for the purpose of considering methods in a more generalized way by seeking to axiomatize their foundations. But let it be confessed that axiomatic systems are in their own right a source of delight.

Any person who has studied geometry is acquainted with an axiomatic system. Many persons, even some who are otherwise mathematically minded, are left by such a study with the mistaken notion that Euclid's theorems are true in the same way it is true that the Pacific is the largest ocean on earth. The only truth a theorem in any axiomatic system possesses, however, is the kind of truth invested in $q$ when it is implied by $p$ and $p$ is asserted. A theorem is said to be true because the prior theorems are themselves so spoken of. These prior theorems, of course, ultimately rest on the assertions of the axioms. And what kind of assertions are these? They rest on nothing — least of all their self-evidence, to which some high school texts used to appeal. Another way of saying that they rest on nothing is to say they are postulated. They are of the order of assumptions: "Let these axioms be assumed and we shall investigate what will follow from them."

The misconception about the truth of Euclidean theorems is inseparable from that about the self-evidence of the axioms. Especially since Euclidean geometry is so completely reliable in surveying, for instance, it seems to follow that its theorems — for that matter its postulates — must be true in the empirical sense, *i.e.*, in the sense that the Pacific is the largest ocean is true. In short, if the Euclidean postulates are only the 'blackboard' assumptions that mathematicians and logicians insist they are, how did their implications ever get off the blackboard into the plat book? The answer is a bold one: even the simplest arithmetic is a blackboard system. And the fact that the grocer charges his customer fifty cents for three cans of soup and a dollar for six requires the same accounting for. To make matters worse, it is historically the case that men were using the arithmetical operations for millennia before Peano undertook to axiomatize arithmetic.

The explanation reconciling cash registers to Peano's five axioms is this: the abstract system Peano concerned himself with is a pure abstraction (or pure 'blackboard'). It is widely *applicable* to real life because our empirical world is popping out all over with the only characteristic required by the blackboard postulates. The latter deal with discrete units, and that is what cans of soup and almost everything else in real life consists of. The case is the same with Euclid. His abstract system can be expressed without any reference to points, lines, and planes. Even when these very terms are used, it must be remembered that none of them occupies any physical space. Because the lines light travels in our atmosphere fit the primitive idea of line, and the earth's surface fits the primitive idea of plane, and the center of X-marks on the tops of pegs fits the primitive idea of points, etc., surveying can use the implications of the abstract system.

The reader has a right to ask what these remarks, however interesting, have to do with symbolic logic. A great deal! Logicians, when they axiomatize, have one eye on the *applicability* of a system to ordinary 'logical' thought processes and discourse. This they refer to as the *semantic* aspect of a system, and in this connection they speak of the *interpretation* of a system. Human discourse is the only semantic meaning, but of course it is only one of various theoretically possible interpretations (*cf.* the application to circuitry). But quite apart from any interpretation — very consciously and deliberately apart from it — logicians try to make sure that the system can stand uninterpreted as a pure abstraction. The widespread and century-old muddle-headedness about the 'self-evidence' of axioms is the reason they are so much on guard. Thus carefully separated from any linguistic interpretation, the system is contemplated syntactically. And so considered, the question of what the symbols may mean is set aside in favor of the question of how they are to be manipulated. Were Euclid writing today, he might in this vein decline to refer to points, lines, and planes until after his system had been adduced without reference to them, thus circumventing any irrelevant appeal to self-evidence.[7]

It is not surprising, then, that axiomatic systems are very rigorously adduced. They generally start with a group of primitive ideas. In the propositional calculus these would be the letters $p$, $q$, $r$, etc., forgetting as best one can that these will ever be allowed to stand for language propositions. Included in the primitive ideas is that of operators affecting the variables.

Because the number of primitive operators should for the sake of elegance be as few as possible, it is typical in such systems that negation and conjunction are primitive and the other operators are defined in terms of them. Negation and disjunction can also be a primitive minimum. So can negation and implication. Sheffer's stroke function (see page 7), which in effect

[7]Carl G. Hempel cites two Euclidean proofs which depend in part on assumptions which are not a part of the postulates — assumptions unconsciously admitted into the system when it is interpreted in terms of points, lines, and planes but whose extraneousness can be seen once the system is contemplated apart from any interpretation. See his "Geometry and Empirical Science," *American Mathematical Monthly*, LII (1945), 7–17, particularly the second section.

combines negation and disjunction, has been used as a sole primitive operator:

$$\bar{p} \text{ is defined as } p \mid p$$
$$p \vee q \text{ is defined as } p \mid p . \mid q \mid q$$
$$pq \text{ is defined as } p \mid q . \mid p \mid q$$

By defining the other conventional operators in this way, the system can be enriched without invoking any additional primitive ideas. An alternative more commonly resorted to is to confine one system, designated as the object system, strictly to the primitive operators. The other operators introduced by definition are then used only for making abbreviated statements representing more cumbersome ones in the object language.

In the above definitions, note the occurrence of dots as punctuation. Obviously these, or parentheses, or some punctuating convention, must be adduced as a primitive symbol in connection with another fundamental idea, that of beneformation (*cf.* Section 5 of Chapter I). For until it is made perfectly clear what collections of symbols qualify as unambiguous, no domain for the system is marked off, *i.e.*, there is no way of knowing what is and what is not a statement in the language being organized.

Once the primitive symbols are in order, the axioms are next. The axioms for the propositional calculus set forth in Whitehead and Russell's *Principia Mathematica*[8] are these:

(1) $p \vee p . \supset p$ (3) $p \vee q . \supset q \vee p$

(2) $q \supset . p \vee q$ (4) $p \vee . q \vee r : \supset q \vee . p \vee r$[9]

(5) $q \supset r \supset : p \vee q . \supset p \vee r$

Once these assumptions are set down ("primitive propositions" they are called in the *Principia*), it is time to produce theorems implied by them. Here again a rigorous syntactical approach is appropriate, because it is important that the theorems follow by rule rather than by any intuition founded on the semantic interpretation. A typical rule of inference is unique and runs something like this: If the horseshoe connects two well-formed formulas as a statement of the system, and the antecedent is a statement of the system, then the consequent is a statement of the system. At this point most systems specify exactly what constitutes a proof; in this regard systematic development is like the development of a deductive method (*cf.* page 44), which is to be expected inasmuch as the system must incorporate some method of making deductions in order to produce its theorems.

It might seem that the theorems are all to follow solely by this rule, but this would overlook the role played by $p$, $q$, etc., as variables, *i.e.*, as capable

[8]Originally published 1910–13. The axioms referred to appear on pages 96ff. of the second edition (Cambridge: The University Press, 1950). Other axioms listed there would in this account be called primitive ideas.

[9]The wedges are dotted, naturally, because it is their associativity that the axiom is asserting.

of representing other expressions — negations, disjunctions, implications, or even quite complex well-formed formulas. In effect there are two ways of writing lines in a proof: (1) by this substitution of other well-formed formulas for the variables occurring in the axioms, a line is written which is not considered a deduction from an axiom but a repetition of it or, more exactly yet, an instance of it;[10] (2) by virtue of the rule of inference mentioned in the previous paragraph.

### EXERCISE V–3

1. a. Which schema of the Deductive Method in this text does the second axiom of the *Principia* most resemble?
   b. Which axiom most suggests Rule II of the Method?
2. Here is the way the *Principia* derives its first theorem ($p \supset \bar{p} \,.\!\supset \bar{p}$):

   Substituting $\bar{p}$ for $p$ in Axiom 1, we assert: $\bar{p} \vee \bar{p} \,.\!\supset \bar{p}$

   By the definition of implication ($p \supset q$ was defined as signifying $\bar{p} \vee q$) we assert from the previous assertion: $p \supset \bar{p} \,.\!\supset \bar{p}$

   Imitating this style, see if you can derive these other early theorems without stepping outside these axioms, the two rules mentioned at the end of this section, and the above definition:

   a. $q \supset .\, p \supset q$
   b. $q \supset r \supset:\, p \supset q \supset .\, p \supset r$
   c. $p \supset .\, p \vee p$
   *d. $p \supset p$ (This proof begins with a substitution on the theorem proved in *b* of this question, uses the first axiom and the theorem derived in *c*.)
3. McGinnis remarked, "I don't see why these axioms are called that. I can deduce every one of them by the Deductive Method of Chapter II." Write a brief clarification of this point.

## 4. Analyticity and Completeness in the Propositional Logic

There are several desiderata for any such logical system as that described in the previous section. One is that the axioms, like the primitive ideas, be as few as possible. In 1926, Bernays showed that axiom 4 of the above list need not be entered as an axiom but can be deduced from the other axioms.[11]

[10] Recall that note 5, p. 29, remarked that the use of schemata "underlies all symbolic logic."

[11] Paul Bernays, "Axiomatische Untersuchung des Aussagenkalkuls der Principia Mathematica," *Mathematische Zeitschrift*, Vol. 25, 1926.

This streamlining repaired an inelegance in the original system.[12] More recently methods have been developed for doing what at first sounds like an impossible task, that of demonstrating that an axiom *cannot* be derived from other axioms and is therefore independent. These proofs are delightfully ingenious. The principle to which they commonly appeal will be returned to later.

Although the preface of this volume contrasts methods with systems, in many respects any good method resembles a system. (Notice that the phrase *Deductive System* throughout this book refers to what is properly a method, but which must bear certain systematic features. To avoid confusion in the present context, it will now be referred to as our Deductive Method.) Economy is of the essence of a system. In a deductive method economy of rules is sacrificed for convenience; rules overlap and are redundant but are left so because they are handier for making inferences than a rigorously economical set would be. What corresponds to the axioms of a system appears in diverse forms in the Method of this text. The schemata for Rule II are to some effects axioms. In a way, GP permits resort to three axioms (the specified schemata).

An important point of similarity between systems and methods is that both aspire to be *analytic*. For a method this means that it is incapable of proving something which is not valid. For a system it means that every theorem, or thesis, generated from the axioms must be logically true. These two are plainly the same thing. Every theorem of an analytic system is a tautology, and every argument which an analytic method is capable of demonstrating is also a tautology. To show that a system for the propositional logic is analytic is fairly simple. By means of the truth table (the independent, *i.e.*, *extra-systemic*, means of identifying a tautology) each of the axioms can be tested and seen to be tautologous. This is to say that the original statements from which all others of the system are to be generated possess the desired tautologous quality. If a non-tautologous statement is to be introduced into the system at any point, it must be introduced by means of one of the rules. But the rule of substitution yields only new formulations of the same previous statements, so that their tautologous quality is preserved. And as for the rule of inference, it can be shown (again by the truth table) that it cannot generate a non-tautologous statement unless at least one of the two statements it uses (the implication and its antecedent) is a non-tautology. It follows that because there is no first non-tautologous statement in the system, it is analytic.

The Deductive Method in this book can similarly be shown to be analytic by examining each of its four rules.[13] Rule I resorts only to certain schemata which (by the truth table) can each be shown to be an equivalence, so that

[12]The *Principia* has since been retouched in numberless ways, Quine's *A System of Logistic* (Willard Van Orman Quine, Cambridge: Harvard University Press, 1934) being one of the most important. These many-sided improvements do not diminish the importance of the monumental original, but only testify thereto.

[13]This passage limits itself to the propositional logic.

any line justified by Rule I must write a statement which is not only implied but which is actually the equivalent of some previous statement in the proof. Rule III (apart from its use in connection with Rule II) only reasserts what has already been asserted. This is true even of the conjunctions of previous statements which it authorizes, since each line of a proof stands asserted and all lines are in effect conjoined (*cf.* the exposition of this point in Section II–8). Rule II uses only certain schemata which can be tested individually by the truth table to verify that what they allege to be implied really is implied without regard to the truth value of individual propositions in the implicans. Rule II itself requires particular attention. The schemata it exploits may be said to have the general form of $PQ \supset R$, wherein for schema

| | | |
|---|---|---|
| $a$, | $P$ stands for $p$, $Q$ for $p$, | and $R$ for $p \vee q$; |
| for $b$, | $P$ stands for $p$, $Q$ for $q$, | and $R$ for $p$; |
| for $c$, | $P$ stands for $p$, $Q$ for $p \supset q$, | and $R$ for $q$; |
| for $c'$, | $P$ stands for $p$, $Q$ for $\bar{p} \vee q$, | and $R$ for $q$; |
| for $c''$, | $P$ stands for $p$, $Q$ for $\bar{q} \supset \bar{p}$, | and $R$ for $q$. |

Rule II authorizes the replacement of $P$ (or $Q$) in any expression wherein it stands under "only horseshoes to its left, dots, or wedges" by $R$ provided $R$ could be deduced from an asserted $P$ (or $Q$), *i.e.*, provided the other expression of the implicans (the one corresponding to $Q$ (or $P$)) be asserted. Can it be shown that this license results only in expressions which are really implied?

By the rule's wording two of the five operators admitted to the method are eliminated out of hand from those which may stand over $P$. Any horseshoe standing over it is required to be to its left; this fact means that such an expression can be replaced (by TH) by an equivalent one in which the horseshoe gives place to a wedge without affecting $P$. Accordingly, the rule has the effect of limiting $P$ to standing under wedges and dots only. The question now becomes this: If $PQ \supset R$, then is it true that, given $Q$, $PS$ implies $RS$, and that $P \vee S$ implies $R \vee S$, and that $PS \vee T$ implies $RS \supset T$, etc.? A rigorous demonstration that this is the case must be extended in principle to cover the *etc.* Although this is not difficult, it will not be included here. But the questions prior to the *etc.* are readily symbolized:

Is $PQ \supset R \supset: PS \cdot Q .\supset RS$ tautologous?
Is $PQ \supset R \supset: P \vee S \cdot Q .\supset R \vee S$ tautologous?
Is $PQ \supset R \supset: PS \vee T \cdot Q .\supset RS \vee T$ tautologous?

If these are answered affirmatively, then Rule II authorizes only legitimate inferences. And of course the truth table is the independent warranty that these statements are tautologous.[14]

[14]It may appear that this omits consideration of the horseshoes or other operators over which $S$ may stand. But $S$ in these symbolized formulations of the question need not be considered as only a variable; it can as well stand for a complex expression made up of any operators whatever.

Rule IV, finally, authorizes the introduction of tautologies into the deduction. Can this possibly result in deducing something which does not follow? A symbolized formulation of this question is equally easy. Let $P$ represent the conjunction of premises of an argument, and $Q$ the conclusion. The argument is then symbolized as $P \supset Q$. Let there be conjoined to the premises a tautology, $R \vee \overline{R}$, representing the GP that is introduced. The argument so modified can be expressed as $P \cdot R \vee \overline{R} .\supset Q$. By the truth table this is seen to be equivalent to the first formulation, $P \supset Q$, from which it follows that the introduction of the GP has in no way affected what $P$ really does imply. In summary, because none of the rules of the method serves to introduce a line which does not follow, any conclusion reached from given premises (including the special case of the null premise) must be valid, since such a conclusion is a line of the proof.

The next desideratum of an axiomatic system is that it be complete. There are several meanings of *completeness;* two are enough for this brief survey. If a system's symbolism is adequate for expressing every statement possible in its domain, it is said to be functionally complete. The statements possible in the domain of the propositional logic can be defined in terms of the truth table: whatever various combinations of true lines and false lines are possible in two variables — these are the statements possible in two variables. There are sixteen of these combinations (*cf.* note, page 22). When three variables are considered 256 distinct statements are possible. For $n$ variables, the statements possible are 2 raised to the $2^n$ power. These statements are all that are possible. Except for the statement in which all lines are false (this can be expressed by $p\overline{p}$), a Boolean expansion can express any of these possibilities; hence, the wedge, conjunction, and negation are enough by way of operators. Any of the combinations of primitive operators mentioned earlier either include or can generate these operators and more, so that these systems are complete in this way. Obviously, the operators used in the Deductive Method suffice for this completeness there.

Another kind of completeness is less trivial. A propositional system is said to be deductively complete if it is capable of generating all tautologies as theorems. In this respect completeness complements analyticity. If every thesis is a tautology, a system is analytic; if every tautology is a thesis, it is complete. The source of the term is plain: the development of a system — the generation of its theses — is completed whenever it can be shown that the theses already developed are sufficient for proving anything that can be a thesis, *i.e.*, any tautology. Suppose that the theorems already developed include the transformations used in our Deductive Method under Rule I. These are sufficient (as was shown in Sections II–3 and II–4) for reducing any statement in the domain to a DNF — even to a simple DNF; and because they are all equivalences, the transformations can be worked in reverse, which is to say that given a simple DNF equivalent to a statement, the statement can be yielded. Because any statement that is a tautology is equivalent to the simple DNF $p \vee \bar{p}$, and because this particular statement soon appears as a theorem (it is number 10 in the *Principia*) in those systems in which it is not adduced as an axiom, any system which has reached the point described can be regarded as capable of generating every tautology, *i.e.*, as complete. In short, by proving $p \vee \bar{p}$ and the theorems constituting the needed transformations, every tautology is rendered provable.

A method can be said to be complete if every valid argument can be proved by it. There are but three general ways in which an argument can be valid: (1) the conclusion is tautologous (this subtends the case of the null premise proof), (2) the premises are self-contradictory, and (3) both are contingent but the premises imply the conclusion (this subtends the case in which premises and conclusion are equivalent). The first kind can be proved in our Method by starting with a GP, $p \vee \bar{p}$, expanding it to the required number of variables, then transforming this DNF into the tautology to be proved. This would hardly be the shortest way of writing such a proof, but we are concerned here with specifying some way which is in principle certain to effect the proof. Arguments of the second kind can be proved by conjoining the premises, reducing this conjunction to DNF which, because it is self-contradictory, can then be used (resorting to schema *a* under Rule II) to prove any conclusion, including the one required. For an argument of the third kind, let the conjoined premises be transformed into DNF and then expanded to its Boolean expression, *i.e.*, a disjunct for each true line of the truth table. Suppose we call this line $n$ of the proof. By using like transformations and working backward from the final line (the conclusion), a Boolean expansion equivalent to the conclusion can be written. Let this line be $n + 1$. The transformations latterly involved are all equivalences, so that the backward-working is convertible into forward-working, *i.e.*, the lines deduced from the conclusion can be used to go from line $n + 1$ *to* the conclusion. Line $n$ being equivalent to the premises conjoined and line $n + 1$ being equivalent to the conclusion, the question remaining is whether the Deductive Method is capable of deducing line $n + 1$ from line $n$. On the prior hypothesis that the argument is valid, line $n + 1$ must have either just the same disjuncts as line $n$, in which case the proof is completed by striking one of the two lines, or else those same disjuncts plus one or more others, in which case it is deducible from line $n$ by schema *a* of Rule II. Since every valid argument can be proved, then, the Deductive Method is complete.

If every tautology is a thesis of a complete system and no non-tautology is part of an analytic system, any effective means of determining whether a given statement is tautologous or not becomes a *decision procedure*, *i.e.*, a means of deciding whether the statement is a thesis of the system. The *decision problem* is that of producing a decision procedure. The propositional logic abounds in decision procedures. The truth table is one. The use of trapezoids is another form of the same method. The cross-out technique is another method.

## EXERCISE V–4

1. a. The next to last paragraph of this section speaks of writing line $n$ as a Boolean expansion and $n + 1$ in the same way, in which case an implied line $n + 1$ might have more disjuncts. Were line $n$ and the (implied) line $n + 1$ both in simple DNF, in what *two* ways might $n + 1$ differ from $n$?

   b. What two schemata could be appealed to in deducing line $n + 1$ in such a case?

2. Janet believes there is no warrant for asserting that everything proved by the Deductive Method is a tautology. "Many of the lines in the proofs we wrote in the propositional logic were contingent," she protests. "Why, some were even self-contradictory!" Write a paragraph clarifying this matter.

## 5. The Independence of Axioms

The means of demonstrating the independence of the axioms of a system can now be delineated. In the discussion of analyticity it was remarked that the tautologous character of the original axioms is passed on by the rule of inference or, stated another way, tautologousness is said to be a characteristic *inherited* from the previous statements when the rule of inference is used. A characteristic having no relation to tautologousness (or, for that matter, to contingency or to self-contradiction) can nonetheless be similarly inherited. Such characteristics and their inheritability underlie the proofs of the independence of axioms. In these proofs, the inherited feature is nothing more than some tabulatable property (similar to *1* or to *0* in the truth table, but lacking in meaning) that can be shown to be passed along in the deduction only if it occurs in the antecedent statement. Then, if by consulting the table (which has been designed for this purpose) all but one of the axioms can be shown to have this characteristic and that one is shown to lack it, it follows that the axiom cannot be deduced from the others. The characteristic is generally a value from a multi-valued table quite artificial apart from the purpose at hand. But it makes no difference what the value stands for, or how the table is made up, or which value is lit upon. If the above conditions obtain, the independence is established.

## 6. Analyticity and Completeness in the Quantificational Logic

The quantificational logic is more complicated than the propositional, as the reader now knows well enough. The complication consists essentially in the fact that $p$ can be *simply* true or false whereas the truth of $x\ Fx$ depends on a host of specific propositions, $FaFbFc \ldots$, $a$, $b$, $c$, etc., being elements in the universe. $\exists x\, Fx$ is equivalent to $Fa \vee Fb \vee Fc \vee \ldots Fn$, $n$ being the number of elements in the universe. Worse yet, the number of disjuncts in such an expression can easily exceed the number of elements. Even an expression as simple as $x\, \exists y\, Fxy$, which becomes $Faa \vee Fab \vee Fac \vee \ldots Fan \cdot Fba \vee Fbb \vee Fbc \vee \ldots Fbn \cdot \ldots$, will require $n^2$ disjuncts when so expressed. Whereas a statement in the propositional logic is logically true if it is tautologous, *i.e.*, if the truth table shows it to be true under every truth assignment of its component variables, a logically true statement in the quantificational logic must obtain in any non-empty universe no matter what the number of elements contemplated. A self-contradiction in this logic must be logically false in every non-empty universe. A contingent statement is true of some universes, false with respect to others.

Metalogical considerations are accordingly more complex here. A system is analytic if every thesis thereof is true of every non-empty universe, and complete if every thesis thus logically true can be proved. Axioms and rules of the simpler logic form a part of those required by a system in the more complex. Similarly, the rules adduced by a method in the propositional logic are extended to form a method in the quantificational.

Suppose that a self-contradictory conjunction of quantified statements were restated with respect to each element in the universe in the manner appearing in the first paragraph. There would then result self-contradictory statements respecting some element(s) in the universe, *i.e.*, a self-contradiction expressed in the propositional logic. Such a way of attacking deduction would be sound in theory but hopeless in practice. The tremendous leverage afforded by instantiation consists in its obviating such a clumsy reduction to the simpler logic by permitting us instead to single out the element in which the self-contradiction obtains. For example, in these statements,

1. $\exists x\, Fx$
2. $x(Fx \supset Gx)$

$\sim$C: $x\, \overline{G}x$

instantiation permits this kind of reasoning: of the things declared by the first statement to be $F$, let one be singled out and designated as $w$. It makes no difference which of the disjuncts of $Fa \vee Fb \vee Fc \vee \ldots$ be the one(s) that renders this disjunction true; the element it mentions will be referred to by this representative letter, $w$. By instantiating the second statement also to $w$, we are in effect discarding from the conjunction $Fa \supset Ga \cdot Fb \supset Gb \cdot Fc \supset Gc \cdot \ldots$ every conjunct except the one that matters — the one touching the element mentioned in the first statement. The same discarding characterizes the instantiation of the third statement to $\overline{G}w$. The propositional self-contradiction is now at hand.

A further leverage provided by such instantiation is that it enables us to disregard the question of how many elements there may be in the universe. The above reasoning obtains whether there be one or one million elements: the reduction by instantiation has shown that in *any* universe the statements are self-contradictory.

It will be instructive to sketch out proofs of analyticity and completeness for one of the methods of this text for the quantificational logic. To speak more accurately, the method to be considered is somewhat *like* those of this text. It can be described as one which:

1. uses no generalization (by foregoing these, the seven rules listed on pages 113f can be reduced to the four restricting instantiations — those appearing on page 139),
2. reveals a self-contradiction in the negation of the statement (or argument) in order to establish the logical truth of the original, and
3. does not include the notion of identity (this exclusion is not indispensable, but because the purpose here is to give the gist of such proofs rather than achieve their results, the present sketch will serve as well with this omission).

(Nowhere in this text was the quantificational logic explicitly given a standard of beneformation. This was implicitly done by allowing two additions to the language of the propositional logic: (1) propositional functions with arguments duly affixed (*cf.* page 78, note 2) were admitted to figure in the language in the same way that propositional variables do, and (2) quantifiers, existential and universal, were admitted as operators affecting arguments falling within their scope and so constituting compound statements to figure like any other compound statements.)

The reliance on instantiations for reducing quantified statements to propositional ones, once construed as a shortcut for the clumsier method of referring to every element in the universe, will be seen to yield a self-contradiction in the simpler logic if and only if the conjunction of quantified statements is false of every universe. For this reduction which refers to every element, being the equivalent[15] of the quantified statements, will contain a self-contradiction if and only if they do; and instantiation is nothing more than a selection of certain pertinent elements in which the self-contradiction can be shown to be located. So the contradiction revealed by instantiations, if it is to be a spurious one, must rest on fallacious instantiation. Hence, the task of showing analyticity is that of showing that no illicit instantiation is possible if the four rules are observed. Completeness will depend on showing that the rules prevent no licit instantiations. That is, the proof must show that every licit instantiation is unrestricted and every illicit one prevented.

If an illicit instantiation is one which violates the rules, then of course the previous sentence is involved in gross circularities. *Illicit* must be given a meaning independent of the rules, and then the rules must be shown to mechanically repudiate the illicit and admit all that is licit. *Illicit* will presently be so defined by marking out a generic fallacy that is unique. Then all the possible ways of instantiating will be listed. Those committing the generic fallacy will be found to be prohibited by the rules; those not committing the

[15] *Equivalent* is used here to mean *makes the same statement in another way* (*cf.* the first paragraph of this section).

fallacy (and thereby licit, the fallacy being unique) will be found to be unprevented by the rules.

*This generic fallacy is the unwarranted assumption that two possibly distinct elements of the universe are one.* Suppose that some instantiation has been made to $w$, for example. Any subsequent instantiation must be either to that same representative or to another. If it is to another, no error can thereby be committed even if in the universe in question the element represented by $w$ should happen to be the same as that now represented, let us say, by $y$; all that results is that the element has two designations: the instantiations do not allege that $w$ and $y$ are distinct. If, on the other hand, the subsequent instantiation is again to $w$, the instantiations allege that the elements are one and the same. If this really is the case, no error is involved, but if the elements are possibly distinct, the second instantiation is illicit. This limitation to two general cases, mutually exclusive and exhaustive, is the warrant for regarding this fallacy as generic and unique.

Now as to the possible sequences of instantiations:

1. All instantiations made to fresh (*i.e.*, not previously used) representatives are therefore licit. Such instantiations may not be useful for demonstrating the self-contradiction sought for, but they will be licit. No rule prohibits such instantiations.

2. An EI made to a representative from a previous (or simultaneous) EI is one form of the generic fallacy; $\exists x\ Fx$ and $\exists x\ Gx$ may refer to two distinct elements, and an instantiation to the same representative assumes them to be the same. Rule 1 prevents this. Because such an instantiation is always subject to the fallacy mentioned, Rule 1 does not prevent any licit instantiation.

3. An EI made to a representative previously (or simultaneously) UI'd

   a. when the UI'd representative determines the EI'd one. This is illicit, being another specific form of the generic fallacy: to instantiate $x\ \exists y\ Fxy$ to $Fww$ is to assume that the $y$ to which $x$ is $F$ is the same element as $x$, whereas the implicans would allow of their being distinct. Rule 7 prevents this. The bookkeeping procedure by which determinations are kept track of and which the latter part of Rule 7 presupposes extends the prohibition to mediated determinations which, being determinations still, are equally illicit.

   (At this point Rule 6 should be considered. The two kinds of instantiations outlawed by this rule can be schematized as shown here.

| | |
|---|---|
| 1. $x\ \exists y\ Fxy$ | Premise |
| 2. $x\ Fxy$ | 1 EI |
| 1. $x\ \exists y\ Fxy$ | Premise |
| 2. $\exists y\ Fxy$ | 1 UI |

   Although questions can be raised about the meaning of the lines deduced, they are really beside the point. It is enough that they obscure a dependence and can thereby give rise to fallacies. The question that must be answered is whether Rule 6, while guarding against a *procedure* which can entail error, can come to bar a licit and needed instantiation.

Inasmuch as the method here contemplated must eventually instantiate every variable, all that Rule 6 can be charged with (considering that it requires certain instantiations to be made simultaneously) is that it hastens these instantiations. This is to say that any licit instantiations falling under its scope can be reached by simultaneous instantiation as well as by any slower route. Rule 6 is simply a device for insuring that Rule 7 is not carelessly circumvented: it prescribes, in effect, that a dependence must appear in the table of EIs when it no longer appears in the lines of the proof.)

b. when the UI'd representative does not determine the EI'd one. This is licit. The following deduction may appear to commit the generic fallacy, but it does not.

| | |
|---|---|
| 1. $x\ Fx$ | Premise |
| 2. $\exists x\ Gx$ | Premise |
| 3. $Fw$ | 1 UI |
| 4. $Gw$ | 2 EI |

The $w$ of line 3, $F$ being predicated in line 1 of every element, can be any element in the universe. This $w$ 'represents' each element without as yet designating any element in particular. Only when the EI of line 4 singles out the (possibly unique) element declared by line 2 to be $G$ and designates it by $w$, does $w$ come to specify a particular element. In fine, were the EI to be effected before the UI, no problem would exist. The present explanation is needed only to show that the order of the two instantiations is really indifferent. This licit instantiation is not prevented by any rule.

4. An EI made to a singular plainly commits the generic fallacy for it assumes that the element declared by the implicans of the EI to have a certain predicate is the same element as that indicated by the singular. This is prevented by Rule 4.

5. A UI made to a representative from a previous (or simultaneous) EI
   a. when resulting in a violation of Rule 7, must be illicit by the reasoning in 3a above.
   b. when no determination is involved, is licit. The rules offer no obstruction to this instantiation.

   That is, what is prevented is illicit; what is not prevented is licit.

6. A UI made to a representative from a previous (or simultaneous) UI. Always licit, since what is affirmed of everything can be affirmed of any particular element. No rule prevents such instantiations.

7. A UI made to a singular is licit by the reasoning under 6, and is unprevented.

This list of instantiations is exhaustive. Under each heading it is pointed out that the illicit instantiation is prevented by the rules and the licit ones are not.

This is equivalent to saying that the rules prevent all illicit instantiations — this constitutes the analyticity of the method, and prevents no licit ones — this constitutes its completeness.

(If from the fact that every licit instantiation is possible it followed that the right instantiations, *i.e.*, those which would evoke the self-contradiction contained in the negation of a valid argument, could necessarily be hit upon, then proofs involving polyadic functions would be as mechanical as those limited to singly quantified monadic functions. But once polyadic functions are admitted, there results a domain in which proofs are no longer necessarily mechanical. And because no general means other than proof exists for determining whether an argument of this sort is logically true, no decision procedure exists in this domain. That is, there is nothing to play the role here which the truth tables or trapezoids play in detecting tautologies in the propositional logic or which trapezoids play in the domain of singly quantified monadic functions.[16] Indeed, it has been shown by Church that no such procedure applicable throughout the quantificational logic can exist.[17] It is the case that some special, as opposed to general, decision procedures have been discovered.[18])

Throughout the above treatment it has been presupposed that in making instantiations, one representative will be recorded as depending on another whenever, and only when, this really is the case. The expression 'really is the case' is used to call back to mind the distinction made in Section 6 of Chapter IV. There, after explaining the nature of dependence, we subsequently adopted a working definition of dependence (page 108) to guide us when entering notations in the table of EIs. That definition requires that we construe one variable (referred to as $y$) as dependent on another ($x$) if (a) $y$ is existentially quantified within the scope of the universal quantification of $x$, and (b) either $x$ itself or some dependent thereon occurs within the scope of $y$. What now must be shown — lest the above proof be worthless — is that it is possible to free any statement of spurious dependences, *i.e.*, from such dependences as meet the test of the definition but do not really involve

[16] $x\ \exists y(Fx \equiv Gy)$, although multiply quantified, can be rewritten as a singly quantified statement: $\exists x\, Fx \supset \exists x\, Gx \cdot \exists x\, \overline{F}x \supset \exists x\, \overline{G}x$. Because every multiple quantification of monadic functions can be thus reduced to an equivalent expression that is singly quantified, the trapezoid method is really adequate for such expressions as well. Strictly speaking, then, the domain to which the trapezoid method is limited is that of *first order* monadic functions. *First order* demarcates quantificational systems in which only the arguments of functions are quantified. *Second order* systems are those in which the functions as well are quantified. This text is confined to the first order calculus, but the second can easily be exemplified by symbolizing the statement that *any function* that has the properties of transitivity and symmetry is also reflexive

$$F(x\ y\ z(FxyFyz \supset Fxz)\ x\ y(Fxy \supset Fyx) \supset x\ y(Fxy \supset Fxx))$$

Here the universal quantifier $F$ binds each occurrence of $F$ within its matrix.

Second order monadic functions, by the way, also have a decision procedure.

[17] Alonzo Church, "A Note on the *Entscheidungsproblem*," *Journal of Symbolic Logic*, Vol. 1, 1936, pp. 40–41.

[18] W. Ackermann, *Solvable Cases of the Decision Problem* (Amsterdam: North-Holland Publishing Company, 1954).

any shifting. This will also satisfy a pledge implicitly given when that definition was first introduced.

This part of the proof will go forward along these lines: TQS and TQD will be briefly examined because they will prove useful to the ensuing explanations; the concept of minimal scope of a quantifier will be introduced; then it will be shown that in a minimally quantified expression any dependences falling within the definition are genuine.

Let it first be noted that TQD need never be resorted to in the contemplated method simply because whatever TQD can do can be done without it. Just as the schemata of TQD were deduced independently (*cf.* the first question in the Exercise on page 120), it would be possible to get along without those schemata by deducing, in place of a needed schema, the very line it would permit one to write.

By temporarily stepping back into the more general (seven-rule) Deductive Method, it is easy to see that TQS is similarly dispensable. Whatever a TQS does can be done by two lines without it: the first line following the implicans will be an instantiation of all its variables, the second line (the one reached directly by TQS) will then be justified as a generalization on the first. Condition 1 governing TQS insures only that this imagined substitute process will be correctly carried out as to the rebinding of whatever was bound in the implicans. Condition 3 can be set aside by supposing that curls, horseshoes, and equivalences are to be transformed away. Condition 2 preserves the same order of quantifiers unless both implicans and new line show the variables concerned to be independent. But in this case the interpolated line of instantiations will be unhindered by a dependence so that the generalizing second line can licitly change the order of the quantifiers.

But even this reliance on the seven-rule method can be dispensed with. In what follows, TQS can be limited severely. Condition 3 will be set aside as above. Condition 1 will need no justification. And as to condition 2, there will be no need to change the relative position of any unlike quantifiers by TQS as long as either is within the scope of the other![19] Once their scopes exclude each other, then of course they are related only by disjunction or conjunction, both of which operators are commutative.

A quantifier will be said to have minimal scope when neither TQS nor TQD will avail to reduce its scope further. An expression will be said to be minimally quantified if each of its quantifiers is of minimal scope. The process of transforming any given expression into a minimally quantified equivalent can start with some quantifier whose ranking as an operator is lowest, or as low as that of any other quantifier. Any number of like quantifiers standing immediately over the same matrix should be construed as being of equal rank. The choice among them will be determined by convenience only.

[19]To make this good may require in practice a great deal of rearranging of the order of disjuncts and conjuncts, but the concern at this point is with theory rather than practice.

The means of reducing a quantifier's scope will depend on the major operator within its scope:

a. If the quantifier be a universal (let it be referred to as $x$) and
   1. the matrix be a disjunction, use TQS to reduce the scope of $x$ to only such disjuncts as contain $x$. If now there be no disjuncts that are conjunctions, the scope is minimal. If one or more disjuncts be conjunctions, convert the matrix to CNF and use TQD to distribute the quantifier $x$ over each conjunct.
   2. the matrix be a conjunction, use TQS to reduce the scope of $x$ to only such conjuncts as contain $x$. Then use TQD to distribute the quantifier $x$ over the remaining conjuncts.
b. If the quantifier be an existential (call it $y$) and
   1. the matrix be a disjunction, use TQS to reduce the scope of $y$ to only such disjuncts as contain $y$. Then use TQD to distribute the quantifier $y$ over the remaining disjuncts.
   2. the matrix be a conjunction, use TQS to reduce the scope of $y$ to only such conjuncts as contain $y$. If now there be no conjuncts that are disjunctions, the scope is minimal. If one or more conjuncts be disjunctions, convert the matrix to DNF and use TQD to distribute the quantifier $y$ over each disjunct.

Repeated applications of 1 and 2 will result in a minimal scope which consists of either a single function (or quantification) or else a series of two or more such single functions (or quantifications) *disjoined*, if the quantifier in question be universal; *conjoined*, if it be existential.

When each quantifier has been so reduced, the resulting minimally quantified expression will be equivalent to the original expression, since only Rule I was used in the process. The remaining question is whether in such an expression any dependence-by-definition can be spurious. That it cannot will be shown by examining a minimally quantified expression in which the relationship between an existential and a universal standing over it is as tenuous as can be contrived. If no spurious dependence appears in it, then *a fortiori* other such expressions will be free of them.

For this purpose consider a matrix standing under the universal, $w$, (1) which is given three disjuncts for reasons that will appear presently.

$$w(Fw \vee Gw \vee Hw) \tag{1}$$
$$\exists y(w(Fwy \vee Gw \vee Hw)Jy) \tag{2}$$
$$\exists z(\exists y(w(Fwy \vee Gwz \vee Hw)Jy)Kz) \tag{3}$$
$$x(\exists z(\exists y(w(Fwy \vee Gwz \vee Hw)Jy)KzLxz) \vee Mx) \tag{4}$$

In each of the disjuncts in (1) $w$ must appear; otherwise its scope is not minimal. Next the existential quantifier $\exists y$ is prefixed: (2). Let whatever else lies within the scope of $\exists y$ be represented by $Jy$ which can be any sort of conjunction, however complex, in each conjunct of which $y$ occurs. This $Jy$ is not germane to the present problem; it is appended to show that the formula being developed is a generalized one. Because $y$ must also appear somewhere in the first conjunct (the quantifier $w$ and its scope), we introduce it into one function therein — the $F$ function, which now becomes $Fwy$. This function may well include other arguments quantified by lower or higher ranking quantifiers than those appearing here, but it will at least include $w$ and $y$. To this let $\exists z$ be prefixed: (3). $Kz$ will stand for whatever else may lie within the scope of $\exists z$. The $z$ must be included in the matrix of $w$, but to keep its relation to $y$ as tenuous as may be, let it be the argument of a distinct disjunct, $Gwz$. In (4), the final formulation, $Mx$ stands for further disjuncts the quantifier $x$ may subsume, while $Lxz$ is introduced as one of the functions of $z$ (detached from the previous $Kz$) which renders $z$ dependent on $x$ by the definition. The dependence of $z$ on $x$ is established outside the matrix of $w$ to see if such removal is enough to make spurious the dependence of $y$ on $x$ now indicated by the definition. The proof is concluded by indicating a universe in which $y$ must shift with $x$ (proving that the dependence is genuine) — a universe which, fortunately, can consist of only two elements, $a$ and $b$:

| | | | | | | |
|---|---|---|---|---|---|---|
| $\overline{Ma}$ | $Laa$ | $Ka$ | $Ja$ | $\overline{Ha}$ | $\overline{G}aa$ | $Faa$ |
| $\overline{Mb}$ | $\overline{L}ab$ | $Kb$ | $Jb$ | $\overline{Hb}$ | $Gab$ | $\overline{F}ab$ |
| | $\overline{L}ba$ | | | | $Gba$ | $\overline{F}ba$ |
| | $Lbb$ | | | | $\overline{G}bb$ | $Fbb$ |

This way of establishing completeness is peculiar to the method it applies to; *i.e.*, in a method which relies for its controls on discerning dependences, the proof need show only that the dependences are genuine. For this purpose a minimally quantified expression is a useful theoretical device. For methods referred to in Section 6 of Chapter IV as commonly in use, the proof often takes an opposite direction. There, because any existential within the scope of a universal is suspected, so to speak, of a dependence, it is necessary to seek out an equivalent statement in which no existential stands under a universal. To this end any statement to be proved is put (for theoretical purposes) into one parenthesis with all its quantifiers standing over it (call this statement $A$). A method due to Skolem is then used to write another one-parenthesis statement in which all the existential quantifiers stand to the left of any universal ones; call this statement, in 'Skolem normal form', $B$. Then it is shown that if $A$ be a tautology, $B$ will be its equivalent. Finally, it is shown that if $B$ is provable, so is $A$.[20] And since any tautology in the form of $B$ can be shown to be provable, it follows that any tautology, no matter in what form, can be proved.

The reason minimally quantified expressions (or, alternatively, Skolem normal forms) are to be found in metalogical proofs when the methods the proofs are about get along very nicely without them is that it is easier to

[20] $B$ is equivalent to $A$ only if $A$ is tautologous. The minimally quantified expression is the equivalent of the statement of which it is the transformation whether that be tautologous or not.

demonstrate the reliability of some procedures than others. The procedure for the metalogical proof is chosen because it is easily shown to be reliable, however cumbersome it might be in practice. (Recall that the proof of completeness in the propositional logic proposed to reduce the conjunction of premises and the conclusion to their Boolean expansions — a method one would rarely, if ever, use in writing a proof.) Here are two proofs which, although far from economical, illustrate the procedures which this section demonstrates to be adequate.

| | | |
|---|---|---|
| 1. | $x\ \exists y(FxGy)\quad \therefore\ \exists y\ x(FxGy)$ | |
| 2. | $x(Fx\ \exists yGy)$ | 1 TQS |
| 3. | $xFx\ \exists yGy$ | 2 TQS |
| 4. | $C \vee y\ \exists x(\overline{F}x \vee \overline{G}y)$ | GP |
| 5. | $C \vee y(\overline{G}y \vee \exists x\overline{F}x)$ | 4 TQS |
| 6. | $C \vee y\overline{G}y \vee \exists x\overline{F}x$ | 5 TQS |
| 7. | $\exists y\ x(FxGy)$ | 3,6 c′ |

| | | $x\ w$ |
|---|---|---|
| 1. | $x\ \exists y(FxGy)\quad \therefore\ \exists y\ x(FxGy)$ | EI: $x_3y_4z_8w_9$ |
| 2. | $xFx\ \exists yGy \vee \exists x\overline{F}x \vee y\overline{G}y$ | GP |
| 3. | $xFx\ \exists yGy \vee \overline{F}x \vee \overline{G}y$ | 2 UI EI |
| 4. | $Fx\ Gy$ | 1 UI EI |
| 5. | $xFx\ \exists yGy$ | 4,3 c′ |
| 6. | $C \vee y\ \exists x(\overline{F}x \vee \overline{G}y)$ | GP |
| 7. | $\exists x\overline{F}x \vee y\overline{G}y \vee xFx\ \exists yGy$ | GP |
| 8. | $C \vee \overline{F}z \vee \overline{G}w$ | 6 UI EI |
| 9. | $\exists x\overline{F}x \vee y\overline{G}y \vee FzGw$ | 7 UI EI |
| 10. | $C \vee \exists x\overline{F}x \vee y\overline{G}y$ | 8,9 c′ |
| 11. | $\exists y\ x(FxGy)$ | 5,10 c′ |

The first proof uses TQS (TQD is not needed) in the limited way referred to earlier which is committed to retaining the order of unlike quantifiers as long as one stands over the other. The second follows the same general procedure but obviates TQSs by resorting to instantiations, using four lines to replace each pair of TQSs (lines 2 to 5 and 7 to 10).

## 7. Notes on Monadic Functions

The procedure described in the previous section for reaching a minimally quantified expression will obviously serve for reducing a multiply quantified matrix of monadic functions to a formula of singly quantified matrices. Indeed, an expression is often freed of all dependences on the way to becoming minimally quantified. So to use such a method in practice would be much like burning down the house to cook the meat.

It seems particularly frustrating that there is not some ready means of distinguishing genuine from artificial dependences within the realm of monadic functions. And as soon as one undertakes to perfect some plausible criterion, he is frustrated even further because he is likely to discover that some genuine dependence lies beyond, or some spurious dependence within, its bounds. It would appear, for instance, that if $y$ is a genuine dependent of $x$, the DNF of the matrix will contain some two disjuncts in each of which appears a function of $x$ not found with like sign in the other disjunct and a function of $y$ not found with like sign in the other disjunct — *e.g.*, in the matrix at the end of the next paragraph $Ax$ and $Cx$ are the functions of $x$ satisfying this requirement, and $Py$ and $\overline{P}y$ are the functions of $y$ satisfying it. But this criterion is by itself inadequate.

The following will give an idea of the complexities introduced by proliferating the quantifiers of a matrix. The expression $x\,w(AxBw \vee CxDw)$ has the following singly quantified equivalent:

$$x(AxBx) \vee x(BxDx \cdot Ax \vee Cx) \vee x(AxCx \cdot Bx \vee Dx) \vee x(CxDx).$$

The expression $x\,w\,\exists y(AxBwPy \vee CxDw\overline{P}y)$ can be true only in one or another of the four universes defined in the previous symbolism. In none of these four need the $y$ be dependent on both $x$ and $w$; yet the prefix $w\,\exists y\,x$ is not deducible, *i.e.*, it is not equivalent in the universe defined by the second disjunct, $x(BxDx \cdot Ax \vee Cx)$, nor will $x\,\exists y\,w$ serve as an equivalent prefix in the universe defined as $x(AxCx \cdot Bx \vee Dx)$. The most that can be deduced as an equivalent of $x\,w\,\exists y(AxBwPy \vee CxDw\overline{P}y)$, if the matrix is to be preserved, is:

$$x\,\exists y\,w(AxBwPy \vee CxDw\overline{P}y) \vee w\,\exists y\,x(AxBwPy \vee CxDw\overline{P}y)$$

Recourse to minimally quantified expressions will show the equivalence of the Tim-and-Tom statements (page 121), and demonstrate that $x\,\exists y(Fx \equiv Gy\,.\!\vee \overline{F}y)$ amounts to $\exists y(Gy \vee \overline{F}y)$. But here again, the practical manipulation of monadic matrices having unlike quantifiers is enormously aided by expanding the schemata of TS, not in number but in applicability, so that, for example, $xFx \vee \exists x\,Fx$ can be TS'd to $\exists x\,Fx$, and $x\,\exists y(FxGx \vee Fy \vee \overline{F}xHy)$ directly to $\exists x(Fx \vee Hx)$.

## SUMMARY

Identities expand the power and flexibility of the quantificational logic, accommodating among other things the definite description. Identities are readily incorporated into both the Deductive Method and cross-outs.

Enthymematic (unstated) premises are frequently factual in content, frequently linguistic, *i.e.*, having to do with the property of some function or the meaning of some word. Neither transitivity nor its counterpart, intransitivity, necessarily characterizes a function. The same is true of the properties called symmetry and reflexivity. Simple ways of introducing these formal properties as premises use generalized formulas which assert the property with respect to a particular function.

Axiomatic systems have their semantic aspects but are given a rigorous syntax to avoid the interference of intuition. They contain no empirical truth. A thesis of a logical system is true if it follows regularly from the axioms. Axioms are, finally, assumptions. Economy, whether of operators, rules, or axioms, distinguishes systems from methods.

A system is analytic if whatever it can prove is logically true, *i.e.*, if true-by-being-deduced matches true-by-the-truth-table. It is complete if it can prove whatever is true-by-the-truth-table. Typical proofs of analyticity and completeness for systems in the propositional logic are broadly outlined. The Deductive Method developed in Chapter II is shown to be analytic and complete.

The means by which one axiom of a system can be shown to be independent of the others is sketched in Section 5.

The more complicated problem of proving analyticity and completeness in the quantificational logic is surveyed. Decision procedures are limited in this logic. A cut-down method much resembling cross-outs is delineated and shown to be incapable of affording a fallacious instantiation (and capable of every licit instantiation) by defining generically a fallacious instantiation and then exhaustively considering possible sequences of instantiations. The definition of dependence — critical to the Deductive Method of this text — is then shown to coincide with that of the shifting variable once an expression is minimally quantified. Because minimally quantifying is an effective procedure, this completes the proof of analyticity and completeness of the method under consideration. This completeness proof is contrasted with that for other methods.

The final section attempts to throw some light on a much-neglected area of quantificational logic — monadic functions. Some of the difficulties and challenges within this domain are indicated.

# APPENDIX

In this appendix we shall treat a few dispensable but interesting topics which depend in part on the concepts in Chapter V but would have been an interruption if discussed there. Despite some diversity, they center on one final method — a way of making deductions in the first-order quantificational logic while leaving all the variables quantified. The method of working with bound variables (IV-4) will now be given a more rigorous form.

The features of this method and the topics to be treated incidental to its exposition are these:

1. The method involves neither generalization nor instantiation, save for EG of singulars and UI to singulars.
2. Beneformation peculiar to such a calculus will be formally defined.[1]
3. The development will rest on primitive ideas and rules, one axiom, and derived theorems.
4. The notion of duality will be treated.
5. The calculus will be shown to be complete and analytic.

Starting with this last topic will afford a general survey of what is to follow.

## Proving a Quantified Tautology

(The following steps are illustrated by the Adams-Bennett problem found immediately following Step 4.)

1. $S$ being the tautology to be proved, let it be put in prenex form with its matrix converted to DNF. Call this matrix $M_d$. $S$ in the example is

$$\exists x(Sx\ y(Ty \supset Ryx))\ Ta\ Tb \supset \exists x(SxRaxRbx)$$

$M_d$ is:

$$\overline{S}x \vee Ty\overline{R}yx \vee \overline{T}a \vee \overline{T}b \vee SzRazRbz$$

[1]The domain so defined is of some interest from the standpoint of theory, since every symbolism within these limits is true or false with respect to each possible universe. As was see in IV-5, unquantified functions such as $Fx$ depend for their meaning on a convenient interpretation. Even if it is given that $Fx$ stands for *x is a farmer*, the symbols remain incomplete as a *statement* apart from some convention attending instantiation. On the other hand, $x\ Fx$ is a genuine proposition rather than merely a proposition-*form*, precisely because it affirms that every element is a farmer.

2. Just how the prenexed $M_d$ is tautologous may not be obvious; but *as far as its predicate letters are concerned*, $M_d$ is tautologous by the propositional logic. It may not be clear which predicate letters are finally to oppose which, *i.e.*, given $\overline{F}$-- and $F$-- and $F$-- (let the hyphens stand for variables), it might be either of the latter two which, regarding the first as $\overline{p}$, should be construed as $p$. Conceivably all three predicates might be redundant to the tautology, in which case no opposition at all is required here; but there is some pairing off of functions which is "correct," — by which $M_d$ is tautologous. We assume for the rest of the proof that we know what these oppositions are. Our purpose, remember, is not to show that our insight is infallible but that provided we have the insight, the tools at hand are sufficient to write the proof.

   The next modification works TSs on $M_d$ in such a way as to insure that each required opposition is between a *distinct pair* of predicates. For example, if $M_d$ is made up in part of predicates corresponding to $\overline{p} \lor pq \lor pr$, let these be TS'd to $\overline{p}\overline{p} \lor pq \lor pr$ so that the $p$ of $pq$ can be paired with one *non-p* and that of $pr$ with the other. Let this matrix be called $M_{dt}$. In the example, $M_{dt}$ is

$$\overline{S}x \lor Ty\overline{R}yx \lor Ty\overline{R}yx \lor \overline{T}a \lor \overline{T}b \lor SzRazRbz$$

3. Next, ignoring the arguments and any singulars preserved from $M_d$, let each opposing pair of predicates in $M_{dt}$ be given arguments peculiar to itself, so that the only occurrence of each variable is in the corresponding argument-places of some two such opposing functions or else in one argument-place of some redundant function. Call this expression $M_{dtv}$. Prefix $M_{dtv}$ with universal quantifiers (for each variable); this tautology becomes line 2 of the contemplated proof. Line 1 is a GP disjoining this tautology and its negation. Line 2 is justified as a TS (Th. 4, below, will warrant this). Lines 1 and 2 of the example illustrate this.

4. Lines 3 to the end of the proof are each justified by one or more of the following:

   a. the transformations of the propositional logic
   b. TQS (the restricted TQS of page 178)
   c. TQ (to be introduced below by a definition)
   d. R (without recourse to Rule II)
   e. the Axiom
   f. Theorems derived from the Axiom (these will include TQD and the use of UI to singulars).

After the primitive ideas, rules, axiom, and theorems of this calculus are adduced, it will be shown that

1. line 2 is the 'strongest' implicans of $S$,

2. the calculus affords every licit inference in its domain, so that the derivation of every $S$ from its appropriate line 2 is possible (completeness), and
3. it affords only licit inferences, so that $S$ can be so derived only if it is tautologous (analyticity).

| | | |
|---|---|---|
| 1. | $xsyzrwt(\bar{S}x \vee Ts\bar{R}yz \vee Tr\bar{R}wt \vee \bar{T}s \vee \bar{T}r \vee SxRyzRwt) \vee \sim$ditto | GP |
| 2. | $xsyzrwt(\bar{S}x \vee Ts\bar{R}yz \vee Tr\bar{R}wt \vee \bar{T}s \vee \bar{T}r \vee SxRyzRwt)$ | 1 TS |
| 3. | $xsr(\bar{S}x \vee Ts\bar{R}sx \vee Tr\bar{R}rx \vee \bar{T}s \vee \bar{T}r \vee SxRsxRrx)$ | 2 Ax |
| 4. | $xsr\exists y\exists z\exists w(\bar{S}x \vee Tz\bar{R}zx \vee Tw\bar{R}wx \vee \bar{T}s \vee \bar{T}r \vee SyRsyRry)$ | 3 Ax |
| 5. | $sr(x(\bar{S}x \vee \exists z(Tz\bar{R}zx) \vee \exists w(Tw\bar{R}wx)) \vee \bar{T}s \vee \bar{T}r \vee \exists y(SyRsyRry))$ | 4 TQS |
| 6. | $sr(\bar{T}s \vee \bar{T}r \vee \exists y(SyRsyRry)) \vee x(\bar{S}x \vee \exists z(Tz\bar{R}zx) \vee \exists w(Tw\bar{R}wx))$ | 5 TQS |
| 7. | 〃 $\exists x(SxRsxRrx)$ 〃 $\exists y(Ty\bar{R}yx) \vee \exists y(Ty\bar{R}yx)$ | 6 R |
| 8. | 〃 〃 $\vee\ x(\bar{S}x \vee \exists y(Ty\bar{R}yx))$ | 7 TS |
| 9. | $\bar{T}a \vee \bar{T}b \vee \exists x(SxRaxRbx) \vee x(\bar{S}x \vee \exists y(Ty\bar{R}yx))$ | 8 UI |
| 10. | $\exists x(Sx\ y(Ty \supset Ryx))\ Ta\ Tb \supset \exists x(SxRaxRbx)$ | 9 TH TQ TM |

## Primitive Ideas

The propositional logic is presupposed as a foundation. To that logic are now added:

1. Predicates, also called functions, each designated by a capital letter followed immediately by one or more arguments, each such argument to be either
   a. a singular (in which case it will be conventionally designated by a, b, c, etc.), or
   b. bound by exactly one quantifier (see below), such arguments to be indicated by the use of $x$, $y$, $z$, $w$, $t$, and other letters from the latter part of the alphabet.

   Such predicates are to be treated as are propositional variables. Like these, they are negated by a bar (over the predicate letter).

2. Two singular operators called quantifiers. Each such operator will contain a single letter from the latter part of the alphabet and will be followed by a parenthesis (with or without a prefixed curl) or by other quantifiers followed by such a parenthesis. It will be said to 'bind' each argument designated by that letter within the scope indicated by that parenthesis. Where this parenthesis would contain but a single function, it is to be omitted, and the scope of the quantifier(s) will consist of that one predicate and its argument(s). Quantifiers are of two varieties:
   a. *Existential quantifiers.* When a letter so used is preceded by an inverted 'E', it will be read "there exists an $x$ (*or* $y$, *or whatever the letter may be*) such that"; *e.g.*, $\exists x\exists y\ Fxy$ is read, "there exists an $x$, there exists a $y$, such that $Fxy$."

b. *Universal quantifiers.* When a letter so used is not so preceded, it is read "for any $x$ (*or* $y$, *etc.*)"; *e.g.*, $xy(GxyHy)$ is read "for any $x$, for any $y$, $Gxy$ and $Hy$."

Such a quantifier stands over each function within its scope and over any quantifiers or curls between itself and such functions. In the following, numbers indicate the ranks of operators:

$$\overset{2}{\sim}\overset{3}{\exists y}\,\overset{4}{x}\,\overset{5}{z}\,Fxy \overset{1}{\vee} \overset{2}{x}\,\overset{3}{\exists z}\overset{4}{\sim}(Fxz \overset{5}{\vee} Fzax \overset{5}{\vee} \overset{6}{\exists y}\,Gy)$$

A quantifier is negated by the use of a single curl placed immediately to its left. The parenthesis ordinarily following the curl is to be omitted in this case. A quantifier so negated is negated by removing that curl.

These two operators can be defined in terms of each other; for this reason either can be construed as primitive. Electing the universal as primitive, we give it this definition:

Df.: $x\,Fx \equiv FaFbFc\ldots$ (for each element of the universe)

The existential quantifier is then defined by fiat, using the universal:

Df.: $\exists x\,Fx \equiv \sim x\,\overline{Fx}$

It will be seen below that these two operators are duals of each other.

3. Beneformation: to the definition of beneformation for the propositional logic (pages 13–14) is now added:

"5. It contains, instead of or in addition to propositional variables and operators

a. such functions as have arguments, each of these arguments being either a singular or bound by exactly one quantifier, and

b. such quantifiers as have at least one such function within their scope (whether it contains any argument bound thereby or not)

and the expression becomes well-formed under headings 1 to 4 above, once

c. each curl immediately preceding a quantifier is disregarded and

d. each quantifier immediately to the left of a parenthesis is construed as a curl and

e. all other quantifiers are disregarded and

f. each function (its arguments disregarded) is construed as a propositional variable."

## A Parenthesis about Duality

The dual of an expression in the propositional logic can be defined as the negation thereof (suppose this to be expressed without an initial curl, *i.e.*, with such a curl transformed away) in which each literal is negated. For examples:

| Given this | (whose curl-less negation is this) | the dual is this |
|---|---|---|
| 1. $p \vee q$ | $\bar{p}\bar{q}$ | $pq$ |
| 2. $pq$ | $\bar{p} \vee \bar{q}$ | $p \vee q$ |
| 3. $p \supset q$ | $p\bar{q}$ | $\bar{p}q$ |
| 4. $p \equiv q$ | $\bar{p} \equiv q$ | $p \equiv \bar{q}$ |
| 5. $\sim(p \equiv qr)$ | $p \equiv qr$ | $\bar{p} \equiv \bar{q}\bar{r}$ |
| 6. $p \equiv q \equiv r$ | $\bar{p} \equiv q \equiv r$ | $p \equiv \bar{q} \equiv \bar{r}$ |
| 7. $p$ | $\bar{p}$ | $p$ |

As these show, some duals are negations of each other (*cf.* the fourth example), some are equivalents (*cf.* the last two examples), and some are neither. Any expression is itself the dual of its dual, since a dual can be obtained either by first negating an expression and then negating those literals, or by first negating each literal and then negating that expression. (By this reversed procedure one can start with any expression in the right column and, regarding the middle column as the first step, arrive at the dual in the left column by the second step.)

As a consequence of the De Morgan Theorem (TM), the dual of any expression whose only operators are conjunctions and disjunctions (and bars) is the *same sequence of literals* in which each disjunction is replaced by a conjunction and vice versa. Thus

$$p \vee \bar{q} \cdot r \;.\!\vee\; \bar{r}\bar{s} \vee q$$

has as its dual:

$$p\bar{q} \vee r \cdot \bar{r} \vee \bar{s} \cdot q$$

Considering conjunction and disjunction as duals of each other, and recalling that the dual of any literal is itself, one perceives that the second expression is the dual, component by component, of the first. The dual of any predominantly dot-and-wedge expression can be so set down, although successive operations may be required.

$$\text{Given:} \quad p \vee . q \supset \bar{p}r : p \vee . q \equiv \bar{r} \cdot \sim(\bar{q} \supset p)$$

$$\text{the dual is:} \quad p \cdot (q \supset \bar{p}r)' \vee . p \cdot (q \equiv \bar{r})' \vee (\sim(\bar{q} \supset p))'$$

$$p \cdot \bar{q} \cdot \bar{p} \vee r \vee : p : q \equiv r \;.\!\vee\; \sim(\bar{q} \supset p)'$$

$$p \cdot \bar{q} \cdot \bar{p} \vee r \vee : p : q \equiv r \;.\!\vee\; \sim(qp)$$

In the intermediate lines, let $(A)'$ stand for the dual of an expression $A$. The third line is obtained by observing that $(A \supset B)'$ is $(\sim A)' \cdot (B)'$ and that $(A \equiv B)'$ is the *equivalence* of $(A)'$ and $(\sim B)'$. $(\sim A)'$ is $\sim(A)'$. But although this illustrates the component-by-component duality of dot-and-wedge expressions, it is otherwise a *tour de force* justified only by the clarification afforded; the shorter way to the dual would be to transform the expression into one of bars, dots, and wedges only and then write its dual in one easy step.[2]

That the two quantifiers are duals can be represented thus:

$$(x\ Fx)' \equiv \exists x(Fx)'$$
$$(\exists x\ Fx)' \equiv x(Fx)'$$

For example, the dual of

$$x(Fx \vee Gx), \text{ which means: } \quad Fa \vee Ga \cdot Fb \vee Gb \cdot \ldots$$

is

$$\exists x(FxGx), \quad \text{which means: } \quad FaGa \vee FbGb \vee \ldots.$$

It is trivial that the dual of a tautology is a self-contradiction. But two aspects of duality are of considerable importance: if $A \supset B$ is a tautology, then it is tautologous that $(B)' \supset (A)'$. This fact affords the principal practical use of duality; *dual proof*, or *proof by duality*, means simply that a second proof follows from one already given by virtue of this feature. Inference *a* of Rule II is deducible from inference *b* in this way, *i.e.*, $pq \supset p$ being true, it follows that $p \supset. p \vee q$. Of course, *b* can be said equally well to follow from *a*. Hence a second important feature, namely, that the *tautologousness* of $A \equiv B$ is equivalent to that of $(A)' \equiv (B)'$. (N.B.: these equivalences hold *only* on the condition that the expressions involved are tautologies; the cases with contingencies are not the same.) These laws of duality will be appealed to below.

## Rules

1. The use of substitution on any schema adduced as a primitive idea, or by the following rules, or postulated, or derived, will be licit and tacit. (*Cf.* note 10 on page 167.)

2. The four rules of the propositional logic (Ch. II) are thus extended or restricted:

    a. Rule I is supplemented by TQS (the restricted version). (As a matter of convenience the use of 'TQ' as a code will be continued; but this is really an appeal to a primitive definition rather than a supplement to Rule I. TQD is derived (Th. 5).)

    b. Rule II will exploit only such schemata as are adduced as primitive,

[2] As Professor W. T. Parry has pointed out, duals of expressions symbolized in trapezoids are very easily reached. The dual of such a symbol is its *complement rotated* (about the middle point of the entire symbol) *through 180 degrees*, with due correction of the slants and the lengths of the lines.

postulated, or derived by a previous theorem.[3]

c. Rule III is extended as on page 102. It will also be used to justify the deletion of a quantifier in whose scope remains no occurrence of the variable bound by it (*e.g.*, in Th. 1, below). It will *not* have tacit resort to Rule II, as on page 48.

d. Rule IV requires no change, not even that appearing on page 102.

3. Proofs: *Normally the first line of each proof will consist of either the Axiom or a GP; each subsequent line will be deduced from its immediate predecessor under 2-a, b,* or *c of the above rules.* In this kind of proof each line is necessarily a tautology, since the first line, as well as the schema underlying each deduction, is tautologous.

As a matter of convenience, recourse will be had to proofs of a schematic, or generalized, nature (*cf.* Th. 4). When the results of such a proof are appealed to, it is because the particular schema required by the appeal can be deduced in the way indicated by, or by a substitution on, the schematic proof. Such specific instances of the schematic proof will conform to the previous definition of a proof.

## The Axiom

*Any expression in which Q is a quantifier implies another (well-formed) like it except that an inserted existential quantifier of lower rank, but standing over the same functions, is allowed to capture any number of the arguments previously bound by Q.*

For convenience, this axiom is now expressed schematically:

a. $x\ Fxx \supset x\ \exists y\ Fxy$

b. $\exists x\ Fxx \supset \exists x\ \exists y\ Fxy$

Corollary:

c. $\exists x\ y\ Fxy \supset \exists x\ Fxx$ (from *a* by duality)

d. $x\ y\ Fxy \supset x\ Fxx$ (from *b* by duality)

Just as the axioms of a propositional calculus can be shown by the truth table to be true for all truth values of their components, this axiom can be shown by means of reduction to be true for any universe: by definition, $x\ Fxx$ means $FaaFbbFcc \ldots Fnn$ which by the propositional logic implies $Faa \vee Fab \vee \ldots Fan \cdot Fba \vee Fbb \vee \ldots Fbn \cdot \ldots Fna \vee Fnb \vee \ldots Fnn$ which defines $x\ \exists y\ Fxy$. Formulation *b* is established in the same way.

[3]It can be shown that the propositional logic is complete without Rule II (and even without Rule III), transformations and a single GP schema ($p \vee \bar{p}$) sufficing to produce any tautology. Accordingly, Rule II could be eliminated from the present calculus. But the resulting complication would tend to divert attention from the more important economies, namely, those proper to the quantificational logic.

## Theorems

Theorem 1: $x\ Fx \supset \exists x\ Fx$

| | | |
|---|---|---|
| Proof: | Ax: | 1. $x\ Fx \supset x\ \exists y\ Fy$[4] |
| | R: | 2. $x\ Fx \supset \exists y\ Fy$ |
| | R: | 3. $x\ Fx \supset \exists x\ Fx$ |

Theorem 2: $\exists x\ y\ Fxy \supset y\ \exists x\ Fxy$

| | | |
|---|---|---|
| Proof: | Ax: | 1. $\exists x\ y\ Fxy \supset \exists x\ y\ \exists z\ Fzy$ |
| | R: | 2. $\exists x\ y\ Fxy \supset y\ \exists z\ Fzy$ |
| | R: | 3. $\exists x\ y\ Fxy \supset y\ \exists x\ Fxy$ |

Theorem 3: $x\ Fx \supset Fa$

| | | |
|---|---|---|
| Proof: | GP: | 1. $\exists x\ \overline{F}x \vee x\ Fx$ |
| | Df: | 2. $\exists x\ \overline{F}x \vee FaFbFc \ldots.$ |
| | TD: | 3. $\exists x\ \overline{F}x \vee Fa \cdot \exists x\ \overline{F}x \vee Fb \cdot \exists x\ \overline{F}x \vee Fc \cdot \ldots.$ |
| | R: | 4. $\exists x\ \overline{F}x \vee Fa$ |
| TQ | TH: | 5. $x\ Fx \supset Fa$ |

Corollary: $Fa \supset \exists x\ Fx$ (by duality)

Theorem 4:

*Any quantified expression whose matrix is self-contradictory by the propositional logic is itself self-contradictory, however quantified.*

To $Qx$, which will represent a quantifier of either kind, let an index number be subjoined to distinguish the several variables. (Note the deletion of quantifiers in line 3 required by beneformation.)

Proof:

GP: 1. $Qx_1Qx_2 \ldots Qx_n(Fx_1x_2 \ldots x_n \cdot \overline{F}x_1x_2 \ldots x_n) \equiv$ ditto

TS: 2 ditto

$$\equiv Qx_1Qx_2 \ldots Qx_n(Fx_1x_2 \ldots x_n \cdot \overline{F}x_1x_2 \ldots x_n . \vee p\overline{p})$$

TS: 3. $Qx_1Qx_2 \ldots Qx_n(Fx_1x_2 \ldots x_n \cdot \overline{F}x_1x_2 \ldots x_n) \equiv p\overline{p}$

The last line symbolizes the theorem.

[4]Here all of the arguments bound by the quantifier $x$ (more exactly, the only argument) have been captured by the new quantifier. The expression is still well-formed. Beneformation requires of each argument that it be bound; but of each quantifier, only that it have some function within its scope.

Corollary:

*Any quantified expression whose matrix is tautologous by the propositional logic is itself tautologous, however quantified.* (by duality)

Theorem 5: $\exists x(Fx \vee Gx) \equiv . \exists xFx \vee \exists xGx$

Proof a):

Ax: 1. $\exists x(Fx \vee Gx) \supset \exists x\, \exists y(Fx \vee Gy)$

TQS: 2. $\exists x(Fx \vee Gx) \supset \exists x(Fx \vee \exists yGy)$

TQS: 3. $\exists x(Fx \vee Gx) \supset . \exists xFx \vee \exists yGy$

R: 4. $\exists x(Fx \vee Gx) \supset . \exists xFx \vee \exists xGx$

Proof b):

TH GP: 1. $\exists xFx \vee \exists xGx . \supset \exists x(Fx \vee Gx) \;:\vee\; \exists xFx \vee \exists xGx \cdot x(\bar{F}x\bar{G}x)$

R: 2. $\exists xFx \vee \exists xGx . \supset \exists x(Fx \vee Gx) \;:\vee\; \exists xFx \vee \exists xGx \cdot y(\bar{F}y\bar{G}y)$

TD: 3. $\exists xFx \vee \exists xGx . \supset \exists x(Fx \vee Gx) \;:\vee\; \exists xFx\; y(\bar{F}y\bar{G}y) \vee \exists xGx\; y(\bar{F}y\bar{G}y)$

TQS: 4. $\exists xFx \vee \exists xGx . \supset \exists x(Fx \vee Gx) \;:\vee\; \exists x(Fx\; y(FyGy)) \vee \exists x(Gx\; y(\bar{F}y\bar{G}y))$

TQS: 5. $\exists xFx \vee \exists xGx . \supset \exists x(Fx \vee Gx) \;:\vee\; \exists x\; y(Fx\bar{F}y\bar{G}y) \vee \exists x\; y(Gx\overline{FyGy})$

Ax: 6. $\exists xFx \vee \exists xGx . \supset \exists x(Fx \vee Gx) \;:\vee\; \exists x(Fx\bar{F}x\bar{G}x) \vee \exists x(Gx\bar{F}x\bar{G}x)$

TS: 7. $\exists xFx \vee \exists xGx . \supset \exists x(Fx \vee Gx)$

Corollary: $x(FxGx) \equiv xFx\; xGx$ (by duality)

Theorem 6: $xFx \vee xGx . \supset x(Fx \vee Gx)$

Proof:

GP: 1. $xFx \vee xGx . \supset xFx \vee xGx$

R: 2. $xFx \vee xGx . \supset xFx \vee yGy$

TQS: 3. $xFx \vee xGx . \supset x(Fx \vee yGy)$

TQS: 4. $xFx \vee xGx . \supset xy(Fx \vee Gy)$

Ax: 5. $xFx \vee xGx . \supset x(Fx \vee Gx)$

Corollary: $\exists x(FxGx) \supset \exists xFx\; \exists xGx$ (by duality)

## Completeness and Analyticity

It was shown earlier how any tautologous statement $S$ in the quantificational logic (first order, identities excluded) could be derived from a suitable matrix $M_{\mathrm{dtv}}$ prenexed by universal quantifiers — line 2 of the proposed proof schema. That this is the strongest implicans of $S$ means simply that any other matrix alike as to functions but quantified with fewer universals or with universals and existentials would offer a more limited range of deducible statements. This follows from these considerations:

1. Any $M_{\mathrm{dtv}}$ having a greater variety of arguments must sacrifice its tautologous nature and cannot be deduced from line 1.
2. Any quantification of $M_{\mathrm{dtv}}$'s variables containing an existential is

thereby weaker than line 2, since line 2 will imply that quantification but cannot be deduced therefrom.

3. Any quantification (of fewer variables) by fewer universals is weaker than line 2 for the same reason.

Completeness and analyticity are shown by tabulating exhaustively the inferences possible within the domain and classifying them into two groups:

1. Those within the system will be licit by virtue of their deduction from the Axiom, which was shown when adduced to be true of every universe.
2. For each illicit inference a counter-example is given.

The possible forms of quantified statements are numbered for easy reference:

1. $x\ y\ Fxy$ (or $y\ x\ Fxy$ by TQS)
2. $x\ Fxx$ (or $y\ Fyy$ by R)
3. $\exists x\ y\ Fxy$
4. $\exists y\ x\ Fxy$
5. $y\ \exists x\ Fxy$
6. $x\ \exists y\ Fxy$
7. $\exists x\ Fxx$ (or $\exists y\ Fyy$ by R)
8. $\exists x\ \exists y\ Fxy$ (or $\exists y\ \exists x\ Fxy$ by TQS)

In the following chart, the parenthesis below each valid inference contains the systemic justification of the inference. The parenthesized number below each invalid inference refers to one of the tables below; the particular table affords a two-member universe as a counter-example for that inference.

| | | | | | | | |
|---|---|---|---|---|---|---|---|
| 1 | $\supset 2$, (Ax-d) | $\supset 3$, (Th 1) | $\supset 4$, (Th 1) | $\supset 5$, (Th 1) | $\supset 6$, (Th 1) | $\supset 7$, (Ax-d Th 1) | $\supset 8$ (Th 1) |
| 2 | $\supset 1$, (4) | $\supset 3$, (4) | $\supset 4$, (4) | $\supset 5$, (Ax-a) | $\supset 6$, (Ax-a) | $\supset 7$, (Th 1) | $\supset 8$ (Ax-b) |
| 3 | $\supset 1$, (1) | $\supset 2$, (1) | $\supset 4$, (1) | $\supset 5$, (Th 2) | $\supset 6$, (1) | $\supset 7$, (Ax-c) | $\supset 8$ (Th 1) |
| 4 | $\supset 1$, (2) | $\supset 2$, (2) | $\supset 3$, (2) | $\supset 5$, (2) | $\supset 6$, (Th 2) | $\supset 7$, (Ax-c) | $\supset 8$ (Th 1) |
| 5 | $\supset 1$, (3) | $\supset 2$, (3) | $\supset 3$, (3) | $\supset 4$, (3) | $\supset 6$, (1) | $\supset 7$, (3) | $\supset 8$ (Th 1) |
| 6 | $\supset 1$, (3) | $\supset 2$, (3) | $\supset 3$, (3) | $\supset 4$, (3) | $\supset 5$, (2) | $\supset 7$, (3) | $\supset 8$ (Th 1) |
| 7 | $\supset 1$, (5) | $\supset 2$, (5) | $\supset 3$, (5) | $\supset 4$, (5) | $\supset 5$, (5) | $\supset 6$, (5) | $\supset 8$ (Ax-b) |

| 8 | $\supset 1,$ | $\supset 2,$ | $\supset 3,$ | $\supset 4,$ | $\supset 5,$ | $\supset 6,$ | $\supset 7$ |
|---|---|---|---|---|---|---|---|
| | (6) | (6) | (6) | (6) | (6) | (6) | (6) |

| Table 1 | Table 2 | Table 3 | Table 4 | Table 5 | Table 6 |
|---|---|---|---|---|---|
| $Faa$ | $Faa$ | $\overline{F}aa$ | $Faa$ | $Faa$ | $\overline{F}aa$ |
| $Fab$ | $\overline{F}ab$ | $Fab$ | $\overline{F}ab$ | $\overline{F}ab$ | $Fab$ |
| $\overline{F}ba$ | $Fba$ | $Fba$ | $\overline{F}ba$ | $\overline{F}ba$ | $\overline{F}ba$ |
| $\overline{F}bb$ | $\overline{F}bb$ | $\overline{F}bb$ | $Fbb$ | $\overline{F}bb$ | $\overline{F}bb$ |

## The Deductive Theorem

The theorem generally known as the Deductive Theorem will make the above calculus more useful. Let $P_1$, $P_2$, etc., stand for the premises of an argument and $C$ for its conclusion.

Theorem:

If $P_1P_2 \ldots P_n \supset C$ be a tautology (call it $A$), then the argument in this form (call it *proof B*):

$$
\begin{array}{l}
1.\ P_1 \\
2.\ P_2 \\
\cdot \\
\cdot \\
\text{n. } P_n \qquad \therefore C
\end{array}
$$

can be proved; and if the latter can be proved, then $A$ is a tautology.

A revision in the definition of a proof (*cf. Rules*) is first required, for otherwise *proof B* cannot even be contemplated. Proofs will now be admitted which contain premises as beginning lines and in which each line thereafter is either a GP or is regularly derived from earlier line(s). In such proofs each derived line (including $C$, obviously) will tautologically follow from previous lines.

As to the first part of the biconditional, let the lines immediately following $n$ of *proof B* constitute a proof of $A$ like that described under *Proving a Quantified Tautology*, above. *Proof B* thus comes to contain the line $A$ and will then be completed thus:

$$
\begin{array}{lll}
\text{(A)} & P_1P_2 \ldots P_n \supset C & \\
 & \overline{P}_1 \vee \overline{P}_2 \vee \ldots \overline{P}_n \vee C & \text{TH} \\
 & P_1P_2 \ldots P_n & 1{,}2 \ldots n \text{ R} \\
\multicolumn{2}{l}{P_1P_2 \ldots P_n\overline{P}_1 \vee P_1P_2 \ldots P_n\overline{P}_2 \vee \ldots P_1P_2 \ldots P_nC} & \text{TD} \\
 & C & \text{TS, R}
\end{array}
$$

As to the second conditional, suppose *proof B* to have been successfully completed (without recourse to $A$). Let whatever lines justify $C$ be thought of as conjoined and this conjunction designated by $C_1$; let all lines appealed to in the justification of the several conjuncts of $C_1$ be considered as parts of a conjunction $C_2$. Clearly $C_1$ or $C_2$ may have a premise as a conjunct (in which case let that premise be retained in successive like conjunctions) and if enough such conjunctions are contemplated, there will eventually be one comprised solely of premises (including any gratuitous premises). Call this $C_m$. Then:

$$P_1P_2 \ldots P_n \cdot GP_1GP_2 \ldots . \supset C_m$$
$$C_m \supset C_{m-1}$$
$$C_{m-1} \supset C_{m-2}$$
$$\vdots$$
$$C_2 \supset C_1$$
$$C_1 \supset C$$

By the truth table it follows from this series that

$$P_1P_2 \ldots P_n \cdot GP_1GP_2 \ldots . \supset C$$

which (again by the truth table) is equivalent to $A$ so that the successful completion of *proof B* implies that $A$ is tautologous.

## The Method in Practice

The sort of proofs made possible by the Deductive Theorem is exemplified by two solutions. The first is the Adams-Bennett problem again:

1. $\exists x(Sx\, y(Ty \supset Ryx))$
2. $Ta\, Tb$ $\quad\therefore \exists x(SxRaxRbx)$
3. $\exists x(Sx\, y(\overline{T}y \vee Ryx)\, y(\overline{T}y \vee Ryx))$ 1 TH TS
4. $\exists x(Sx \cdot \overline{T}a \vee Rax \cdot \overline{T}b \vee Rbx)$ 3 UI
5. $\exists x(Sx \cdot \overline{T}a \vee Rax \cdot \overline{T}b \vee Rbx \cdot Ta \cdot Tb)$ 4,2 R TQS
6. $\exists x(SxTaRaxTbRbx)$ 5 TD
7. $\exists x(SxRaxRbx)TaTb$ 6 TQS
8. $\exists x(SxRaxRbx)$ 7 R

In practice, the method can be further simplified by readmitting the schemata of Rule II, as in the justification of line 12 in the proof below. Indeed, Rule II can even be liberalized by admitting quantifiers to the list of operators under which expressions may stand and still be qualified for an inference, *i.e.*, by allowing inferences to be effected *within matrices;* this last on the condition, of course, that the corresponding argument-places of functions involved (as $\overline{T}a$ and $Ta$ in line 5 above) be occupied by the same letter with both occurrences thereof bound by the same quantifier if it be not a singular.

Line 7 of the proof below exploits this extension. By this means, line 8 of the above proof can be justified by *5 c′* and lines 6 and 7 deleted.

The second proof (problem 8, page 122) achieves still another economy by using the unrestricted TQS of page 118:

| | | |
|---|---|---|
| 1. | $\exists x\ y(\overline{K}yx\ Mx)$ | |
| 2. | $x(Mx \supset y(Kxy \lor Cxy))$ | |
| 3. | $\exists y\ x(\overline{C}xy \cdot \exists w\ \overline{K}wy)$ | $\therefore\ \exists x(Mx\ Cxx) \cdot \exists x\ \overline{M}x$ |
| 4. | $t(\overline{M}t \lor z(Ktz \lor Ctz))$ | 2 R TH |
| 5. | $\exists x\ y\ t\ z(\overline{K}yxMx \cdot \overline{M}t \lor Ktz \lor Ctz)$ | 1,4 TQS |
| 6. | $\exists x(\overline{K}xxMx \cdot \overline{M}x \lor Kxx \lor Cxx)$ | 5 Ax |
| 7. | $\exists x(MxCxx)$ | 6 c′ |
| 8. | $t\ z(Ktz \lor Ctz) \lor \exists x\overline{M}x$ | 4 Ax TQS |
| 9. | $\exists w\ \exists y\ x(\overline{C}xy\overline{K}wy)$ | 3 TQS |
| 10. | $\exists w\ \exists y(\overline{C}wy\overline{K}wy)$ | 9 Ax |
| 11. | $\exists t\ \exists z(\overline{C}tz\overline{K}tz)$ | 10 R |
| 12. | $\exists x\ \overline{M}x$ | 11,8 c′ |
| 13. | $\exists x(Mx\ Cxx) \cdot \exists x\ \overline{M}x$ | 7,12 R |

# ANSWERS TO STARRED QUESTIONS

## I–2

1. h. The same symbolism as for (g). *But,* although it connotes an opposition of ideas not suggested by *and,* still affirms both ideas.

## I–6–A

3. The order in which the columns are filled in should be from the higher numbers to the lower.

5. b. $2^n$ lines

12. c. Equivalence

13. d. Adams is right if the attempt to determine whether the expression is well-formed oscillates between 3 and 4 without ending in either. Brown is right if recourse now to one, now to another, heading eventually settles the matter. Which way is it?

## I–6–B

1. h. Supposing (g) to have been written $\bar{p} \supset q$, this would be the same.
   j. And this would be $p \supset \bar{q}$.
   k. $p \supset . r \supset q$
   l. $q \supset p . \equiv p \vee \bar{q}$
   m. The same as (l).
   n. This states that John's being sick implies that he won't be here ($p \supset \bar{q}$), but it also affirms that he is sick. $p \supset \bar{q} \cdot p$ is right; so is $p\bar{q}$. Are these two expressions equivalent?
   o. $p \supset q$ The fact that the sickness *causes* the absence is not captured; *i.e.,* the deletion of the words *for that reason* would leave the symbolism unaffected. Note that it is not affirmed here that John is sick.
   p. $pq$
   q. $\sim(pq)$
   r. $\bar{p}\bar{q}$

2. c. She roars with laughter and voices her mind if and only if she understands little that is said.
   d. If she is neither pleased nor taking her time, then she doesn't voice her mind.
   e. She fails to roar with laughter if and only if she either sulks or is not both pleased and taking her time.
   h. If she either understands little that is said or is both pleased and roaring with laughter, then her taking her time amounts to her either sulking or quenching her thirst. Note that *if and only if* is a much-used rendition of equivalence, but *is equivalent to* or *amounts to* is also acceptable.

## II–2

15. $pq$: $\quad$ $\bar{p}\bar{q}$:

## II–4

1. d. $\bar{p}r \vee qr \vee p\bar{q}\bar{r}$

4. a. $x$ $\quad r \quad pq$

   f. $x \quad x \quad \bar{p}r \quad x \quad x \quad \bar{q}\bar{r}$

   $\bar{p}\bar{q}$

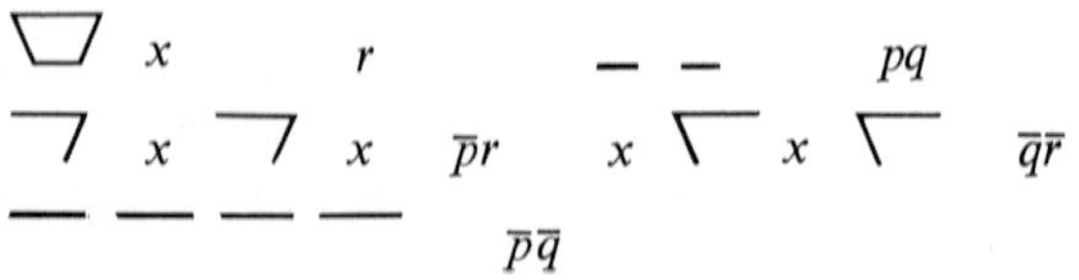

6. Steps 1 and 2, tabulating the Boolean expansion:

| $pqrs$ | $\bar{p}qrs$ | $p\bar{q}rs$ | $\bar{p}\bar{q}rs$ | $pq\bar{r}s$ | $p\bar{q}\bar{r}s$ | $\bar{p}qr\bar{s}$ | $\bar{p}\bar{q}r\bar{s}$ | | |
|---|---|---|---|---|---|---|---|---|---|
| ✓ | ✓ | ✓ | ✓ | | | | | $rs$ | $p'$ |
| ✓ | | ✓ | | ✓ | ✓ | | | $ps$ | $q'$ |
| | ✓ | | ✓ | | | ✓ | ✓ | $\bar{p}r$ | $r'$ |

Development of the prime implicants using trapezoids:

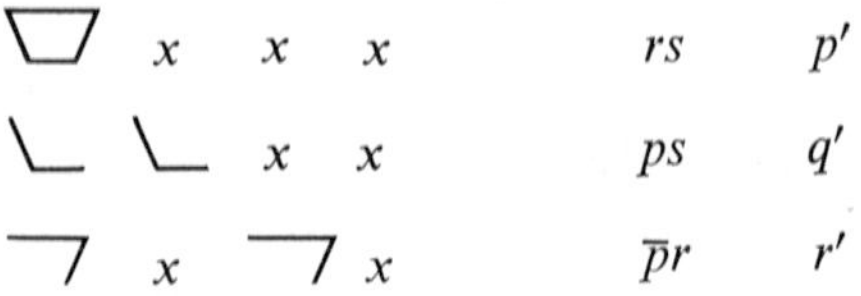

| | | | | | |
|---|---|---|---|---|---|
| x | x | x | | $rs$ | $p'$ |
| | | x | x | $ps$ | $q'$ |
| x | | x | | $\bar{p}r$ | $r'$ |

Step 3, using either the table or trapezoids, yields:

$$p' \vee q' \cdot p' \vee r' \cdot p' \vee q' \cdot p' \vee r' \cdot q' \cdot q' \cdot r' \cdot r'$$

Step 4 (omitting repeated disjuncts in the new DNF) yields $p'q'r' \vee q'r'$. The shorter disjunct ($q'r'$) indicates that $ps \vee \bar{p}r$ is the simple DNF sought.

## II–5–B

3\. 1. $p \vee q$

2\. $q \supset . \, r \vee s \qquad \therefore \overline{p}\overline{s} \supset r$

| | | | | | |
|---|---|---|---|---|---|
| 3. | $\overline{p}\overline{s} \supset \overline{p}\overline{s}$ | GP | 3′ | $p \vee r \vee s$ | 1,2 c |
| 4. | $\overline{p}\overline{s} \supset q\overline{s}$ | 3,1 c′ | 4′ | $\sim(p \vee s) \supset r$ | 3′ TH |
| 5. | $\overline{p}\overline{s} \supset . \, r \vee s \cdot \overline{s}$ | 4,2 c | 5′ | $\overline{p}\overline{s} \supset r$ | 4′ TM |
| 6. | $\overline{p}\overline{s} \supset r$ | 5 c′ | | | |

5\. In the first premise is the wedge over the equivalence or vice versa? Consult the content as well as the structure of the statement. The report appears to deal with number of firemen, alarm system, etc. The equivalence would seem to be a consequent of the report's being right, which is to say that it can be disjoined from the report's being wrong (*cf.* TH).

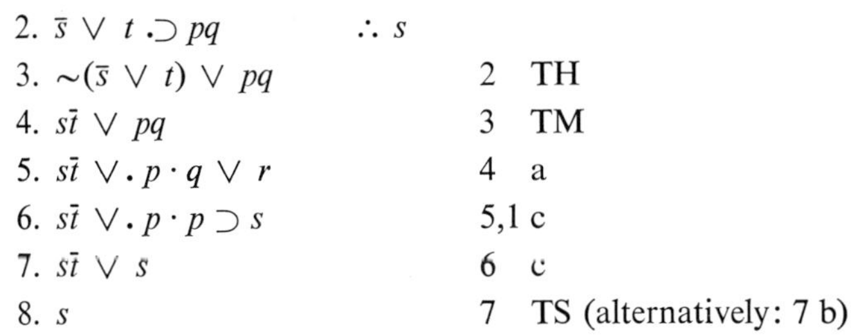

6\. b. 1. $q \vee r . \supset p \supset s$

2\. $\overline{s} \vee t . \supset pq \qquad \therefore s$

3\. $\sim(\overline{s} \vee t) \vee pq$ — 2 TH

4\. $s\overline{t} \vee pq$ — 3 TM

5\. $s\overline{t} \vee . \, p \cdot q \vee r$ — 4 a

6\. $s\overline{t} \vee . \, p \cdot p \supset s$ — 5,1 c

7\. $s\overline{t} \vee s$ — 6 c

8\. $s$ — 7 TS (alternatively: 7 b)

10\. The symbol for the conclusion will always include that for the conjunction of the premises. Under unusual circumstances it can be the same symbol (a special case of including), but ordinarily it will show the same lines *and more; i.e.*, it will be implied by, instead of equivalent to, the conjoined premises. Would the introduction of a GP affect the symbol for the conjunction of the premises?

12\. a. To prove: $pq \vee r . \supset p \vee r \vee s$

1\. $pq \vee r . \supset pq \vee r$ — GP

2\. $pq \vee r . \supset p \vee r$ — 1 b

3\. $pq \vee r . \supset p \vee r \vee s$ — 2 a

13\. c. To test the statement $p \supset q \supset . \, pr \supset qr$

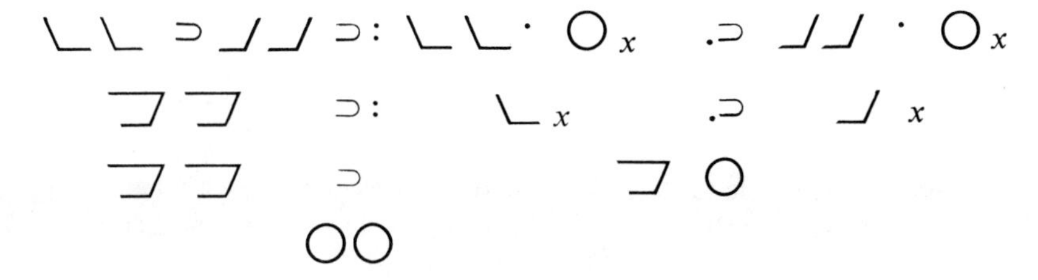

## II–7–B

7. Solutions can of course vary. Here are two solutions such as are called for:

1. $pq \vee \bar{p}\bar{q} \vee \bar{p}q\bar{r}$ b″ b‴

2. $q \vee s \vee pr$ c′

3. $\bar{p} \vee \bar{q} \vee r$ b

4. $p \vee \bar{s} \vee q \vee r$ b′

$\sim$C: $pq\bar{r} \vee \bar{p}\bar{q}\bar{r} \vee \bar{p}qr \vee p\bar{q}r$ a

$/pq\underline{\bar{r}}\,\underline{r}$

LA $/\bar{p}\bar{q}\bar{r}\,\underline{\bar{s}}\,\underline{s}$

$/\bar{p}q\underline{r}\,\bar{p}q\underline{\bar{r}}$

$/p\bar{q}\underline{r}\,\bar{p}q\underline{\bar{r}}$

1. $pq \vee \bar{p}\bar{q} \vee \bar{p}q\bar{r}$ a

2. $\underline{q} \vee s \vee \underline{p}r$ d

LA 3. $\bar{p} \vee \bar{q} \vee r \cdot pq \vee \bar{p}\bar{q} \vee \bar{p}q\bar{r} \cdot \bar{p}\bar{q}\bar{r}\,\underline{\bar{s}}\,\underline{s}$

4. $\underline{p} \vee \bar{s} \vee \underline{q} \vee \underline{r}$ c

$\sim$C: $\underline{pq\bar{r}} \vee \bar{p}\bar{q}\bar{r} \vee \underline{\bar{p}qr} \vee \underline{p\bar{q}r}$ b

## IV–2

4. One in which no *F*s exist.
5. $\exists x(FxGx)$
7. $\exists x\,\bar{F}x$

## IV–6–B

1. a. In deductions as simple as these, the valid solutions will require no particular strategies. A good way of attempting the solution is to write a single instantiated line between premise and conclusion. If that line follows validly from the premise and leads validly to the conclusion, the solution is valid. The solution of (1a) illustrates this:

| | | | |
|---|---|---|---|
| 1. $x\,\exists y\,z\,\exists w\,Fxyzw$ | $\therefore \exists x\,\exists y\,\exists z\,Fxyyz$ | | $x$ |
| 2. $Fxyyz$ | 1 UI EI | | $x\ \ y$ |
| 3. $\exists x\,\exists y\,\exists z\,Fxyyz$ | 2 EG | | EI: $y_2\ z_2$ |
| | | | (Valid) |

Note that in the table of EIs, $x$ and $y$ both stand over $z$ because they are both determinants thereof (the variable in line 1 which was EI'd to $z$ having two universals to its left which were UI'd to $x$ and $y$).

The other EI (to $y$, determined by $x$) could warrant writing $(x)$ above the second entry of $x$ in the EI table. This would certainly not be wrong, since it accords perfectly with the procedure specified in the lines just preceding Rule 7 (page 112); it is omitted here simply as a redundancy, $x$ already being recorded as a determinant of $z$.

## IV–7–B

5. a. The correct writing, using $Pxy$ to mean that $x$ is prior to $y$: *To prove:* $\exists x\, y\, Pxy \vee \exists y\, x\, \overline{P}xy$. If you used a legitimate GP, the proof violates Rule 5, 6, or 7.

7.
| | | |
|---|---|---|
| 1. $\exists x(Sx\, y(Ty \supset Ryx))$ | | |
| 2. $Ta\, Tb$ | $\therefore \exists x(Sx\, Rax\, Rbx)$ | EI: $x_3$ |
| 3. $Sx\, y(Ty \supset Ryx)$ | 1 EI | |
| 4. $Ta \supset Rax$ | 3 UI | |
| 5. $Rax$ | 2,4 c | |
| 6. $Tb \supset Tbx$ | 3 UI | |
| 7. $Rbx$ | 2,6 c | |
| 8. $Sx\, Rax\, Rbx$ | 3,5,7 R | |
| 9. $\exists x(Sx\, Rax\, Rbx)$ | 8 EG | |

10. a. To prove: $x(\exists y\, Fxy \vee y\, \overline{F}xy)$

| | |
|---|---|
| 1. $Fxy \vee \overline{F}xy$ | GP |
| 2. $x(\exists y\, Fxy \vee y\, \overline{F}xy)$ | 1 UG EG |

b. To prove: $\exists x(\exists y\, Fxy \vee y\, \overline{F}xy)$

| | | |
|---|---|---|
| 1. $x\, \exists y\, Fxy \vee \exists x\, y\, \overline{F}xy$ | GP TQ | $w$ |
| 2. $Fwz \vee \exists x\, y\, \overline{F}xy$ | 1 UI EI | EI: $z_2$ |
| 3. $\exists x\, \exists y\, Fxy \vee \exists x\, y\, \overline{F}xy$ | 2 EG | |
| 4. $\exists x(\exists y\, Fxy \vee y\, \overline{F}xy)$ | 3 TQD | |

## IV–8–A

2. a. Every morning Fred shaves.
   b. Some mornings Fred shaves.
   c. Fred never shaves when anyone is looking at him.
   d. The same as (c).
   e. Some persons never look at themselves when they shave.
   f. Fred doesn't shave on those mornings when nobody looks at him.

g. Every pair of persons look at each other on some occasion or other. (What does this statement signify if the same person is chosen for both $x$ and $y$?)

h. The same as (g).

i. There is an occasion when everyone looks at everyone (*or* There is an occasion when every pair of persons look at each other).

3. b. $\exists x(Px\overline{M}x)$

4. a. $\exists x(Px\ y(\overline{O}xy \supset Sxy))$

b. $\exists x(Rx\ \exists y(\overline{S}yWyx))$ and $\sim x(Rx \supset \exists y(SyWxy))$ can each be regarded as correct, although they are not equivalent. Using $Pxy$ to mean $x$ *participates in* $y$, we can also justify this version: $\exists x\ \exists y(RxSyPyx\overline{W}yx)$.

e. $x(Bx \vee Gx\ .\!\supset SxDx)$

f. $\exists x(ExSxc)$

6. Using only these functions: $Jx$ $x$ is a juggler on tour, $Gx$ $x$ engages a good agent, $Wx$ $x$ is well paid; $x(Jx \supset\!.\ Wx \equiv Gx)$. The equivalence must be subordinate to the horseshoe.

With more detailed functions: $Jx$ $x$ is a juggler, $Tx$ $x$ is on tour, $Ax$ $x$ is an agent, $Gx$ $x$ is good. $Exy$ $x$ engages $y$, $Wx$ $x$ is well paid:

$$x(JxTx \supset\!.\ \exists y(AyGyExy) \equiv Wx)$$

Your answer may look different and still be equivalent. If you are unsure about its equivalence, return to this after this chapter is finished.

15. d. $x(Mx\ \exists y(Py\ \exists z(Lz\ w(Uw \supset \overline{M}zw)Eyz)\ \exists t\ Sxty) \supset y(Uy \supset Byx))$

e. $x\ z(Mx\ \exists y(Py\ Lz\ w(Uw \supset \overline{M}zw)Eyz\ \exists t\ Sxty) \supset Tzx)$

18. $\exists x\ \exists y(IxBy\ \exists z\ Sxyz) \supset x(TxWxn \supset y\ z(DyBz \supset \overline{S}yzx))$

## IV–9

3. a. $\overline{F1} \vee p$ is the local quantification of either.

b. $\overline{F1} \vee \overline{G2} \vee Hx1$ is the local quantification of either.

## IV–10

3. 1. $x(Cx \supset Fx)$ $\quad\therefore\ x\ y(CxDyx \supset \exists z(FzDyz))$

2. $C \vee \exists x\ \exists y(CxDyx\ z(\overline{F}z \vee \overline{D}yz))$ GP TQ TM

3. $x(\overline{C}x \vee Fx)$ 1 TH

4. $\overline{C}x \vee Fx$ 3 UI EI: $x_5\ y_5$

5. $C \vee\!.\ CxDyx \cdot \overline{F}x \vee \overline{D}yx$ 2 EI UI

6. $C \vee\!.\ FxDyx \cdot \overline{F}x \vee \overline{D}yx$ 5,4 c′

7. $C \vee FxDyx\overline{D}yx$ 6 c′

8. $x\ y(CxDyx \supset \exists z(FzDyz))$ 7 TS

4. 1. $\exists y\, x(HyFx \vee HyGx)$ $\quad\therefore x(Fx \vee Gx \cdot \exists y\, Hy)$
   2. $C \vee \exists x(\overline{F}x\overline{G}x \vee y\,\overline{H}y)$ GP TQ TM
   3. $HyFx \vee HyGx$ 1 UI EI
   4. $C \vee \overline{F}x\overline{G}x \vee \overline{H}y$ 2 UI EI $\qquad$ EI: $y_3\ x_4$
   5. $Hy \cdot Fx \vee Gx$ 3 TD
   6. $C \vee \overline{F}x\overline{G}x$ 5,4 c′
   7. $C \vee Gx\overline{G}x$ 6,5 c′
   8. $x(Fx \vee Gx \cdot \exists y\, Hy)$ 7 TS

6. a. Assign a letter to a letter
   b. Assign a numeral to two columns (or assign a letter to a letter, or failure to use a new letter when deleting an existential quantifier)
   c. Assign a letter to a letter
   d. Same as (b)

## IV–11–A

4. a. 1, ~C: $\underline{F1y2w}\ \underline{\overline{F}3445}$
$$\begin{array}{ccc} & 1 & \\ 1 & 2 & \\ y & w & \\ 4 & 5 & 1 \\ 2 & & 3 \end{array}$$

   b. 1, ~C: $\underline{F1y2w}\ \underline{\overline{F}3435}$
$$\begin{array}{ccc} & 1 & \\ 1 & 2 & \\ y & w & \\ 4 & 5 & 1 \\ & & 2 \\ & & 3 \end{array}$$

   c. 1, ~C: $\underline{F1y2w}\ \underline{\overline{F}x3x4}$
$$\begin{array}{ccc} & 1 & \\ 1 & 2 & \\ y & w & x \\ 3 & 4 & 1 \\ & & 2 \end{array}$$

   d. 1, ~C: $F1y2w\ \overline{F}3443$
$$\begin{array}{cc} & 1 \\ 1 & 2 \\ y & w \\ 4 & 3 \end{array}$$
   *1* cannot be licitly co-assigned with *3* because it stands over $\underline{w}$ as a determinant.

   e. 1, ~C: $F1y2w\ \overline{F}3434$
$$\begin{array}{ccc} & 1 & \\ 1 & 2 & \\ y & w & \\ 4 & & 1 \\ & & 2 \\ & & 3 \end{array}$$
   *4* cannot be assigned to both $\underline{y}$ and $\underline{w}$.

   f. 1, ~C: $F1y2w\ \overline{F}3453$
$$\begin{array}{ccc} & 1 & \\ 1 & 2 & \\ y & w & \\ 4 & 3 & 2 \\ & & 5 \end{array}$$
   *1* cannot be licitly co-assigned with *3* because it stands over $\underline{w}$.

5. d. Hint: Let *Fxy* mean that *x* is married to *y* and *Gx*, that *x* is single. As a universe, use Al and his wife, Betty. In the second premise let *x* be one and *y* the other. (What makes the second premise true under these circumstances?)

## V–1–A

1. g. The most economical writing is

$$\exists x\ \exists y\ \exists z(RxRyRz\ w(Rw \supset . \ w = x \lor w = y \lor w = z)\ x \neq y)$$

Inclusion of $y \neq z$ or $x \neq z$ is redundant. Inclusion of both eliminates the possibility of there being but two persons. If no non-identity is included, the statement that results is that there are persons in the room but no more than three of them.

h. To construe this as a definite description seems to force the meaning. *The most ambitious student* appears to represent a class rather than an individual, *i.e.*, $Ax$ should signify that $x$ is very ambitious. Hence, $x(SxAx \supset Nx)$. (The superlative becomes simply an *a fortiori:* even the most ambitious student needs rest, so plainly the less ambitious ones . . .)

l. $\exists x(Dx\ y(Dy\ y \neq x \supset Txy)\ x \neq h)$ states that there exists one tallest debater and that he isn't Harry. $\exists x(DxTxh)$ states that there exists one or more debaters taller than Harry. The English appears ambiguous on this point. Either rendition seems justified.

m. $Dt\ x(Px\ x \neq t \supset \overline{D}x)$

3. b. an identity.

6. b. Determinations can occur in non-identities (as in identities). Otherwise this contingency is provable.

## V–1–B

6. Here is a correct writing, using $Txyz$ to mean that $x$ talks to $y$ about $z$. Limit the universe to those humans the argument refers to.

1. $x\ y\ \exists z(Txyz \cdot Pz \lor Nz)$
2. $x(Nx \equiv Px)$
3. $\exists x(Px\ y(Py \supset y = x))$ $\quad \therefore\ y\ \exists z\ x\ Txyz$

7.

1. $\underline{\overline{B}1} \lor Fy1\underline{\overline{F}21} \lor Fy1\ \ 2 = y$ b (Note the distribution)

LA 2. $Fxz\ Fwz\ \underline{x \neq w}\ w \neq z\ z \neq x\ B3\ Fy1\ 2 = y\ Fy1\ \underline{4 = 2}$ $\quad 1$

$\qquad y\ x\ w\ z$

~C: $B3$ a $\qquad 2\ 4\ 1$

R1. $\underline{\overline{B}1} \lor Fy1\underline{\overline{F}41} \lor Fy1\ 4 = \not{y}2$ c $\qquad 3$

9. The writing (limiting universe to those at the party):

1. $x\ y\ Txy \supset Fh$
2. $\overline{F}h\ x\ Txg$
3. $\exists x\ y(\overline{F}y \supset \overline{T}xy)$ $\quad \therefore\ g \neq h\ \exists x\ \exists y\ \overline{T}xy$

a  1. $\overline{T}xy \vee \underline{Fh}$

LA  2. $\overline{F}h\, T1g\, \overline{T}xy\, \underline{\overline{T}w2}\, \underline{T1h}$    $x\ y\ h\ g\ w$

b  3. $\underline{F2} \vee \overline{T}w2$    $3\ 4\ 2\ \ \ 1$

c $\sim$C: ~~$g = h$~~ $\vee\ \underline{T34}$

$T1h$

## V–2

1. $x\, y(Fxy \supset \overline{F}xx)$ If you wrote $x\,\overline{F}xx$, prove the equivalence of the two answers.

5. a. The writing and the needed premises:

   1. $Bat\ Nt$    $\therefore\ x(Fx \supset Bax)$
   2. $x\, y(NxFy \supset Bxy)$    unstated premise
   3. $x\, y\, z(BxyByz \supset Bxz)$    unstated premise

   g. Limit universe to class officers. This writing will require considerable rewriting of the unstated premise:

   1. $\exists x\ Mx\ \exists x\ Wx$
   2. $\exists x\ \exists y(x \neq y \cdot MxMy \vee WxWy)$
   3. $x(Mx \equiv \overline{W}x)$    unstated premise

   $\therefore\ \exists x\ \exists y\ \exists z(x \neq y\ y \neq z\ z \neq x)$

   Those rewritings can be avoided by this way of expressing the premises:

   1. $\exists x\ Mx\ \exists x\ \overline{M}x$
   2. $\exists x\ \exists y(x \neq y \cdot MxMy \vee \overline{M}x\overline{M}y)$

   The writing which is most faithful to the original offers a more difficult problem:

   $Lxy$ $x$ and $y$ are of like sex

   1. $\exists x\ \exists y\ \overline{L}xy$
   2. $\exists x\ \exists y(x \neq y\ Lxy)$   $\therefore\ \exists x\ \exists y\ \exists z(x \neq y\ y \neq z\ x \neq z)$
   3. $x\, y(\overline{L}xy \supset x \neq y)$    unstated premise
   4. $x\, y(Lxy \supset Lyx)$    unstated premise

   Even the cross-out solution is rather complex. Notice the double modification of two FPs; in each case the second modification is by virtue of the identity asserted in the branch involved.

a 1. $\bar{L}xy$

LA 2. $\underline{w \neq z}\ Lwz\ \bar{L}xy\ 1 \neq 2$ /6 = 7 7 = 8 /5 = 7 $\bar{L}78$ $L43$

b 3. $\underline{L12} \vee 1 \neq 2$

e′ 4. $\underline{\bar{L}34} \vee L43$

c ~C: $\underline{5 = 6} \vee 6 = 7 \vee 5 = 7$

d *R~C: $\underline{1 = 2} \vee$ ~~2~~ $= 8 \vee$ ~~1 = 8~~ 7 L1~~x~~6

d′ **~C: $\underline{1 = 2} \vee$ ~~2 = 8~~ $\vee$ ~~1~~ $= 8$ $\bar{L}$x8 7 7

| x | y | w | z |
|---|---|---|---|
| 1 | 2 | 8 | 7 |
| 5 | 6 | 3 | 4 |

* This rewriting employs some numerals already assigned simply to lessen the proliferation of numerals.

** This line, being only a reproduction of the line above it, is not properly a rewriting in the sense of a fresh instantiation. It serves only to prevent the cross-outs entailed by the second branch of the LA from being confused with those entailed (in the previous line) by the first branch.

## V–3

2. d. The substitutions for the first line are these: $p \vee p$ for $q$, $p$ for $r$.

# Index

## Summary of the Cross-Out Technique in the Propositional Logic

1. Negate the statement or argument.
2. Transform each conjunct into DNF.
3. Place each DNF on a separate line, but conjoin all wedgeless disjuncts to form the LA. If there be no wedgeless disjuncts, make a branched LA.
4. Use the LA to strike contradicting disjuncts on other lines, adding to the LA (or to the branch being exploited) any disjuncts thus rendered wedgeless.

*The statement is tautologous (the argument valid) if and only if each branch of the LA shows a self-contradiction of two literals.* (p. 59)

## Rules and Schemata of the Deductive System

**I. Transformations** Any assertion can be rewritten with either the whole or any constituent expression replaced by its equivalent (p. 44). *I* is the code if the equivalence is in a prior line; otherwise a schema affords the code. The schemata are:

TH (p. 29) $p \supset q \mathbin{.\!\equiv} \bar{p} \vee q$

$\sim(p \supset q) \equiv p\bar{q}$

TE (p. 29) $p \equiv q \equiv\!. \; pq \vee \bar{p}\bar{q}$

$\sim(p \equiv q) \equiv\!. \; p\bar{q} \vee \bar{p}q$

(p. 45n) $p \equiv q \equiv\!. \; p \supset q \cdot q \supset p$

TM (p. 29) $\sim(pq) \equiv\!. \; \bar{p} \vee \bar{q}$

$\sim(p \vee q) \equiv \bar{p}\bar{q}$

TD (p. 30) $p \cdot q \vee r \mathbin{.\!\equiv} pq \vee pr$

$p \vee qr \mathbin{.\!\equiv} p \vee q \cdot p \vee r$

TS (pp. 31–32) $pq \vee p \mathbin{.\!\equiv} p$

$\bar{p}q \vee p \mathbin{.\!\equiv} q \vee p$

$pq \vee \bar{p}r \vee qr \mathbin{.\!\equiv} pq \vee \bar{p}r$

TQ (p. 82) $x\,Fx \equiv \sim\exists x\,\bar{F}x$

$\sim(x\,Fx) \equiv \exists x\,\bar{F}x$

TQD (p. 117) $xFx \cdot xGx \mathbin{.\!\equiv} x(FxGx)$

$\exists xFx \vee \exists xGx \mathbin{.\!\equiv} \exists x(Fx \vee Gx)$

A TQD collecting two quantifiers in both of whose matrices occurs the same *representative* is illicit (p. 120).

TQS must conform to 3 stipulations (p. 118).